TAR CREEK

LARRY G. JOHNSON

TAR CREEK

A History Of The Quapaw Indians, the World's
Largest Lead and Zinc Discovery, and The *Tar
Creek* Superfund Site.

Published by Anvil House Publishers, LLC, 2017
Owasso, OK
www.anvilhousebooks.com

Originally published by Tate Publishing & Enterprises, LLC, Mustang, OK, 2009

Bottom photo on cover: Panoramic view of Picher Oklahoma—2 years old (1917). (Photo courtesy of Baxter Heritage Center & Museum, Baxter Springs, Kansas)

Published in the Unites States of America

ISBN: 978-0-9839716-5-8 (pb)
ISBN: 978-0-9839716-9-6 (hc)
Library of Congress Control Number: 2017903748

1. History / United States / State & Local / Southwest
2. Technology & Engineering / Mining

To Sherryl,
my wife, my love, and my best friend.

ACKNOWLEDGEMENTS

For some, including this author, writing a book is a daunting task and must be somewhat akin to giving birth to a child. The new parents-to-be are initially filled with joy and excitement as they visualize what their creation will look like, what kind of child he or she might be, and the great things that will be accomplished by their offspring. These initial sentiments are soon replaced with morning sickness, the growing discomfort of the mother as she progresses through her gestation, the realization of the great responsibility the parents have assumed, and finally the pains of birth. An author experiences these same reactions while writing a book. There is an initial excitement about a telling a story not previously told. But, if done properly, the book's gestation is often strewn with doubts about the need for such a book, the author's ability to adequately tell the story, and whether people will really read it. Because of the long gestation period for most books, the importance of the continual support and encouragement of the author's family during the period is exceptionally important, and without the considerable encouragement and support over the last five years from my wife, Sherryl, and my two sons, Philip and Curtis, this book would not have been written.

Others have contributed to the completion of the book, including countless librarians and museum personnel. Of particular note are the research librarians at the Tulsa City-County Library and the University of Tulsa McFarlin and Law Libraries, all of whom were invaluable in their assistance in locating the missing pieces in my research. Raymond Hale of the Dobson Museum and Memorial Center in Miami, Oklahoma, and Catie Myers and Larry O'Neal of the Baxter Heritage Center in Baxter Springs, Kansas, gave invaluable assistance in selecting and assembling the photographs used in this book. Dennis Whitley of Whitley Graphics was of great help in improving my graphic illustrations. Special thanks must be extented to my editor, Amanda

Webb Soderberg, for her gentle but persistent prodding that made this a far better book than it would have otherwise been. Any defects or deficiencies in the book are the responsibility of the author and not of these capable people.

A proper acknowledgement would not be complete without recognition of the past and present inhabitants of the Tar Creek area, Indian and non-Indian alike.

I sat on our couch one evening recently and watched a documentary that I had previously recorded from our local Public Broadcast System affiliate. The Independent Lens film was about the Tar Creek Superfund site. The camera recorded scenes of the desolate landscape, rusted trailer homes, dilapidated houses, huge chat piles, and trash-filled vacant lots in and around Picher, Oklahoma. There were also scenes of a number of well-kept but modest homes, a number of small businesses, and a good school facility. I saw parents and children who loved their town and wanted to make it the best. Even with the uncertainty of whether Picher will survive (and now we know it will not), they seemed a happy people.

At the time of the filming, debate was occurring among the town's people about whether or not the town should be bought out by the government and abandoned. Everyone had opinions. Some wanted to be bought out and leave. Others said they weren't selling and would have to be carried out. But whatever their opinion on the buyout, most of them worried about their children's health, loved their town, loved their school and its children, and many had wonderful memories about growing up and living in Picher, Commerce, Quapaw, and the surrounding Tar Creek area.

After the documentary ended, I sat in silence thinking about those people. I was reminded of the times I walked the streets of Picher and other area towns, talked to the people, and climbed the chat hills as I researched this book. I could not help feeling a tremendous admiration for this people…these ordinary Americans. No pretense was made to be what they were not. They and millions like them scattered across the continent made America great. They fought its wars, built it cities, and made a civil society civil, and then looked beyond their own back yard to help others around the world. Most aren't off to protest the latest perceived wrong, save the most recent endangered species, claim victim status over some slight, or be consumed with the latest pop-culture tragedy-comedy emanating from Hollywood. People often mock their values and call them ignorant and out-of-touch. They aren't perfect and wouldn't pretend to be if they knew how. When life hits them with a fast curve ball, they grimace but carry the pain silently, and then take their base. They take their turn, say "please" and "thank you," and salute the flag. They can't help it. It's in their DNA.

They are generally pitied by the intellectuals, government officials, and far-off social-planners that wish to control their lives as opposed to helping them plan their own lives. To some, these people often become the fodder for their latest version of Farm-aid, some other do-good project, or other celebrity cause with which to consume the time of their own meaningless lives.

What these ordinary Americans do is work hard, hold their families together and provide for them as best as they can, respect authority, love God, and protect their country. They are a tough race of survivors that brought us through the perilous twentieth century including the Great Depression, four major wars, and the cold war. In this hour of the self-absorbed, self-indulgent "me" generation, they and others like them are becoming anachronisms. That is the real internal threat to America.

Larry G. Johnson
Owasso, Oklahoma
June 1, 2008

CONTENTS

Introduction—Beginning of a Journey   15

Part I—Origins of the Quapaw   **23**

The Quapaw—The Downstream People   25

The Quapaw—Old Americans or New Americans?   49

The Quapaw—White Man's World   74

Indian Country to Statehood—Life on the Edge   97

Part II—Lead and Zinc Mining

Era in the Tri-State Region   **117**

Wildcatters and Prospectors   118

Boomtown!   134

Brother Lead and Cousin Zinc   146

Into the Pits   156

Mountains of Chaff   175

Hell's Fringe   190

Miners' Health and Labor Strife   206

Pickets, Pick Handles, and Police   215

Part III—Mining Aftermath and the

Quapaws in the Twenty-First Century   **233**

"The Filthiest Town Known This Side Of Hell"   234

The Quapaw—Into the Twenty-first Century   252

Epilogue—Journey's End   264

Addendum—The North American Indian

—Ancient Origins   284

Bibliography   304

End Notes   318

Index   349

TAR CREEK MAPS AND CHARTS

Fig. 1—Migration of the
Dhegiha Sioux from the Ohio Valley   26

Fig. 2—Quapaw Villages—1682   28

Fig. 3—Louisiana Purchase—1803   46

Fig. 4—Areas Occupied by the Quapaw
Indians—1550s(?) to Present—Chart   58

Fig. 5—Areas Occupied by the Quapaw
Indians—1550s(?) to Present—Map   59

Fig. 6—The Quapaws with the Caddoes—1826–1835   63

Fig. 7—Quapaw Cession—Treaty of 1867   77

Fig. 8—Quapaw Agency Indian Tribal Reservations   87

Fig. 9—Indian Territory Lands—1866–1889   101

Fig. 10—Oklahoma Territory Lands—1889–1906   104

Fig. 11—Indian Territory Lands—1889–1906   105

Fig. 12—Tri-State Mining District—Towns   191

Fig. 13—Tar Creek   238

Fig. 14—Man's Migration to the Western Hemisphere   286

Fig. 15—Prehistory Cultural Areas in North America   288

Fig. 16—Prehistory of Man in Eastern North America   290

PREFACE

Best-selling author James Burke has written several books dealing with the dynamics of knowledge, time, rapidity of change, and interdependence of actions and events of the past on the present. Burke observes that we all are part of a dynamic web of change that links each one of us with one another and to everything in the past. Every person and every act causes change to the web. He postulates that occasionally the simplest occurrence or act will have gigantic repercussions hundreds of years later and that other seemingly cataclysmic events may lead to the ordinary. These links with the past may be very direct and have an identifiable cause and effect relationship. However, most acts or events ricochet through time, as a pinball does on its table, bouncing here and there, causing other actions or events that might occur from a few seconds to centuries later.[1] It is when these seemingly independent acts and events of the past are linked with the present that history comes alive for we become a living part of that history.

When this book was conceived, it was intended as a straightforward recounting of the story of the world's greatest discovery of lead and zinc located in the far northeast corner of Oklahoma. I soon realized that this 1915 discovery was strongly linked to prior as well as current events and that those events should be a part of the story. And, true to Burke's proposition of a dynamic web of change and the interconnectedness of the present with the past, the story grew. This growth soon encompassed the Quapaw Indians and their ancient history, the founding of the United States, the Louisiana Purchase, the transition of Oklahoma and Indian Territories to statehood, the discovery and mining of lead and zinc in the Tri-State region, and the making of the oldest and largest environmental Superfund site in America.

This book does not claim to do justice to any of these topics but attempts to tell an interesting story from a perspective not seen before. Parts of the book deal with grand themes and momentous events. Other chapters deal with the mundane and minutia of life in frontier mining camps. As time progresses, some of the events and players in our story rise to prominence and then recede into obscurity. Other seemingly insignificant actions and events create ripples that become waves that become tsunamis that etch a lasting mark on the shores of history.

The discovery of lead and zinc deposits in 1915 on the Quapaw Reservation at Picher, Oklahoma, serves as the focal point of the book. Several events and links to the past preceding the discovery are chronicled. Events and actions from that discovery to the present are also noted. One might visually describe this story as an hourglass with the discovery of lead and zinc at Picher as the skinny neck through which all of the interconnected acts and events preceding the discovery (the top half of the hourglass) are slowly moving toward (causing or impacting) that discovery. The bottom half of the hourglass is literally the fallout from the discovery or, as Burke might put it, the repercussions ninety years later.

INTRODUCTION—BEGINNING OF A JOURNEY

I had heard about the abandoned lead and zinc mines for a number of years. I knew these mines lay north of Miami, Oklahoma, and extended into southeastern Kansas and southwestern, Missouri, but I had never had an occasion to drive into the area. Interstate 44 travels through Tulsa to the northeast and passes Miami on its east side and continues another twenty-five miles in a northeasterly direction toward Joplin, Missouri.

On May 1, 2003, my wife and I left for our son's graduation from college in Springfield, Missouri, scheduled for the next day. Before we left that morning, an article in that morning's newspaper had caught my eye.[2] The annual "Toxic Tour," a nineteen-mile bike ride over both paved and gravel roads through the Picher Mining District and several of the small towns in the area, was to be held on Saturday morning, May 3. The tour, first organized in 1995, has continued each year in hopes of raising the awareness and understanding of the outside world to the ecological disaster that has happened in the area.[3]

We had often made the three-hour trip to Springfield on the Interstate from our home in Owasso, which is near Tulsa, but I had never paid much attention to the area west of the Interstate between Miami and Joplin. Fueled by the morning's newspaper article and a considerable amount of previous media coverage about what had become known as the Tar Creek EPA Superfund site, my curiosity was stirred. I found myself looking to the west as our car sped along the Interstate between Miami and Joplin. Between the trees to the west bordering the Interstate I caught a glimpse of the tops of two or three of the infamous chat piles. How many times had I traveled this road and not noticed them? Perhaps there was a story to be told, a story behind the headlines and details of massive environmental damage, state and congressional action (or inaction, depending on your point of view), and the efforts of a multitude of federal agencies. The "Toxic Tour" was my chance to see the area from the seat of a bicycle and learn more. I knew that I had to go.

We returned to our home late Friday evening. By 7:00 a.m. Saturday morning, I had loaded my bicycle into the back seat and part of the front seat of our Cadillac of ancient vintage and was headed to Miami. Even before I arrived at the small community of Picher after passing through Miami and Commerce, I spotted the large mounds of mining wastes called "chat piles." It was a perfect day for a ride. The blue Oklahoma sky was slightly overcast and the temperature was in the upper sixties.

The city park was not hard to find as US 69 becomes Connell Street, Picher's present-day main street, and heads to the Oklahoma-Kansas border at the north end of town. There were only a few side streets, many with abandoned buildings. To the casual observer, Picher appeared to be a typical decaying small town with one major exception: the huge piles of gray chat in close proximity to homes, streets, businesses, abandoned buildings, and schools. The surreal scene struck me as an enormous sandbox filled with conical shaped mountains towering over miniature houses, streets, and cars. One such mound sets a block west of the city park, its edge immediately at the side of a street towering over a small wooden house facing it from the opposite side. We road past that chat pile when we left the park. The sides of the chat pile rise at almost a forty-five degree angle to a peak well over 100-feet above street level. Another much larger chat pile set a couple of blocks to the east of Connell Street. Yet, I realize from my research that this giant-sized sandbox stretches over forty square miles.[4]

There were between fifteen and twenty people at the city park. Most were teenagers or younger from a local church wearing lime green tee shirts with "Staff" printed on the back. There were three or four adults, organizers of the event, who were directing the youth. I paid the $10 entry fee that entitled me to ride on the tour and receive a water bottle, a neck scarf, a disposable camera, and a tee shirt emblazoned with *Toxic Tour, May 3, 2003, Picher, OK, T.E.A.L*

[Tribal Efforts Against Lead], *S.E.A.L.* [Student Efforts Against Lead]. I also received a plastic tag to be attached to my bicycle. It had blanks for medical information, emergency contacts and phone numbers, blood type, medications, allergies, and other medical information...not encouraging.

There were several police cars present to guard the various highway crossings as the bike-riders progressed along the designated route. Two were BIA (Bureau of Indian Affairs) Police; others were from the Picher Police Department and the Ottawa County Sheriff's office. As 10:00 a.m. approached, only about fifteen riders had registered. I visited with Rebecca Jim, a tour organizer and executive director of LEAD Agency, Inc. We discussed her thoughts on the best way to get rid of the chat piles and the suggestion that the area be turned into a 10,000-acre wet land. She believed the wildlife that would live in the wetland or visit it in their migratory routes would continue to become contaminated and spread the lead poisoning to other areas. It seemed that everyone had an opinion as to what should not happen, but no one had a good solution to deal with polluted ground water and the enormous chat piles.

I knew I was in trouble when I saw the other, younger bikers lift their bikes from bumper racks on the backs of their SUVs, check tire pressures, and fine tune and make adjustments to brakes, gears, and various other attached equipment. Most wore tight-fitting spandex shorts above bulging, ride-hardened calf muscles and loose nylon tops favored by the committed riders who think nothing of doing fifty-mile rides after work. Even a mother and her two pre-teens had the right clothing, bicycles, and helmets. Somehow my jeans, knit shirt, leather shoes, and twenty-six inch, three-speed upright 1960s style bike didn't fit in. The stares at this out-of-shape fifty-something didn't bother me as much as one of the ride organizers who repeated several times that there would be a "draggin wagon" for those that didn't want to (she meant "couldn't") finish the ride. I later accused her of looking at me each time she made the announcement. We were off at 10:00 a.m. with the admonishment that "this wasn't a race." I wondered if that was for my benefit. It was then that

I realized that I was the only one without a helmet. Several blocks from the start I also realized I had forgotten my ball cap, something I would later regret even on this mildly overcast day.

It was obvious that most if not all of the riders were there for the ride and not to learn about Tar Creek. I caught up with most of the riders at the first rest stop just as they were leaving the paved road on the second leg of the journey. Soon I caught up with three or four riders who had stopped to pick up trash and place it in small plastic sacks each had brought with them. Apparently, it was their custom to pick up trash along a section of road on their weekend outings. I assumed this was their small contribution to cleaning up the planet. It seemed a hopeless if somewhat ridiculous task given the considerable amount of trash that had been dumped. A dump-site for abandoned refrigerators, other appliances, bags of trash, and assorted refuse could be found on almost every mile, usually next to a chat pile or the concrete foundations or rusting remains of a pull derrick or milling operation. I made it to the second rest stop on the Kansas-Oklahoma border and was soon joined by the bicycling environmentalists with their trash neatly separated into recyclables and non-recyclables. Duty done, they soon sped off, never to be seen again.

Much of the land not covered by chat piles had been planted in crops. We crossed a swiftly flowing stream that ran a foot deep across a small concrete bridge. Was this Tar Creek flowing into Oklahoma from Kansas? I learned later that it was the stream for which the Superfund site had been named. The water flowing out of Kansas didn't look that bad, but I knew it was laced with chemicals not good for the body. I continued east, crossed Highway 69, and came to several large chat piles on the Oklahoma side of the line that were owned by the Quapaw Indian Tribe. Following a pause at the rest stop next to the chat piles, I continued east, and was now the last rider of the group. The mother and her two children were about a half-mile ahead and seemed to be keeping an eye on me. Occasionally, one of the event organizers would drive by and ask how I was doing. Trying not to look stressed or breathe too hard, I would smile and say, "Just fine." Eventually, the washboard gravel road branched to the south, and I knew I was over half way according to the little map given to us at the start of the tour.

The next rest stop was at Hockerville. Only one building remained of the once bustling community. The sign in front of the small church said, "Welcome—Hockerville Baptist Church," and listed service times. According to the inscription on the stonework above the door, the building once housed the First State Bank. A young lady attending to the rest stop directed me behind the building to a large sinkhole caused by the 1918 cave-in. The large depression in the earth was now overgrown with large trees and brush and had become a decades-old dumping ground for large quantities of trash.

I soon passed the outskirts of the small town of Quapaw. A lone officer driving a BIA Police vehicle waited at the corner and took my picture as I road past. Heading east on a well-paved road, I knew that I would make it to the end. The mother and two children had long since disappeared into the distance, but I found them at the next-to-last rest stop. After a quick drink of water, I left them chatting with the workers dispensing water and refreshments and crossed over Highway 69 a couple of miles or so south of Picher.

Now on a rough gravel road, I soon came to the last rest stop at the point where Tar Creek passes under the Douthat Bridge, named after the long-gone community of the same name whose remains rested behind fences posted with warnings against trespass. This site is known for the orange mine water that spews into Tar Creek just as it flows under the bridge. When mining ceased in 1969, the pumps were turned off and the mines filled with water. The minerals in the mine dissolved and created acidic mine water, which began seeping out of the ground and into Tar Creek in 1979.[5]

I talked with the lady dispensing water and refreshments. She told me that her grandfather was killed in the mines. I moved to the north side of the bridge and watched as the orange water from Lytle Creek merges with the somewhat clearer water of Tar Creek as it flows under the bridge. She told me that down the road just south of Cardin is where the chat pile dust, scraped from 1,600 lawns of area residents, is being taken. Most residents believe the removal of the lead-tainted dust from their yards is blowing back onto their properties.[6]

With legs aching and other parts of my anatomy in distress, I pushed on through Cardin once again and entered Picher. As I neared the park from which I had departed three hours earlier, I was told that four more riders hadn't completed the ride. I'm certain that their absence was due to other distractions, not fatigue. I coasted into the park and dismounted with as much grace as possible. I was detained by one of the young "Staff" members who said that I must answer questions about my ride for a survey. I agreed only after she allowed me to sit (gingerly) on a nearby park bench. Interview completed, I loaded my bike and departed.

I drove north on Connell Street and located the Picher Mining Museum. Louis Hile was the eighty-eight-year-old volunteer who kept the museum open from 1:00 p.m. to 4:00 p.m. daily, Monday through Friday, to accommodate the 400 visitors it received during the spring, summer, and fall. Mr. Hile was the lone occupant, so we sat and visited undisturbed for an hour. He worked in the mines for only a brief period but was an employee of the Cox Machine Shop, one of many that serviced mining equipment. His job frequently took him down in the mines. He had lived in and around Picher all of his life, and his father worked in the mines in the late teens and early 1920s.

Photographs of thousands of faces of miners lined the walls of the museum. Many photographs were group pictures of the various mining company employees. Some were taken in and around the mines as they worked. All lead one to conclude that mining was a rough, hard, and dangerous life in the early part of the twentieth century.

As I guided my car through the tollgate and onto the turnpike, I reflected on the day's events. Yes, there was a story that needed to be told. But there were so many facets to the story—the Indians who owned most of the land; the rich discoveries, the boomtowns, and the hardships and death that were a common denominator of the miner's life; and the massive environmental damage to the forty-plus square miles that was making headlines with regularity. Where should one begin?

PART I—ORIGINS OF THE QUAPAW

The Quapaw people took a circuitous route to their reservation located in the far northeastern corner of Oklahoma that once contained the largest and richest deposits of lead and zinc the world ever knew. "The Quapaw—The Downstream People" recounts the Quapaw's flight down the Ohio River and probable arrival at the Mississippi River in the 1550s, separation from their Siouan kinsmen, and their sojourn in Arkansas to the time of the Louisiana Purchase in 1803.

"The Quapaw—Old Americans or New Americans" covers the period from 1803 to the end of the Civil War in 1865. The momentous events following the dramatic territorial expansion of the United States resulted in two major upheavals in the tribe's history that threatened their survival: removal to Indian Territory in the 1830s and the devastation to the tribe and its reservation caused by the Civil War. This period covers the remainder of their years in Arkansas, brief time in northwestern Louisiana, and ultimate removal to a reservation in Indian Territory. In the land that became Oklahoma, two parallel worlds existed between the end of the Civil War and statehood in 1907—one Indian and the other white. "The Quapaw—White Man's World" tells the story of the tribe's struggle to survive as a tribe. During that period, the Quapaw moved from near extinction as a tribe to rapidly adapting to the white man's world. This transformation greatly influenced their destiny in the twentieth century beginning with the vast discoveries of lead and zinc on their tiny reservation.

"Indian Country to Statehood—Life on the Edge" tells the story of the white settlers from the end of the Civil War to statehood in 1907—a story of outlaws, land runs, and the raucous efforts to found a brand new state. These parallel worlds of the Indian and white settlers would forever intersect with the discovery of huge amounts of mineral wealth beneath the Oklahoma prairie.

To better understand this journey, one must trace the probable migration and subsequent pre-history and history of man in the Western Hemisphere. The pre-history of the Quapaws is examined in the "Addendum—The North American Indian—Ancient Origins" and gives an account of the arrival of the ancestors of the North American Indian, the groups that migrated to and occupied southeastern Canada and the eastern United States, and the Indian wars that drove the Quapaw from their Ohio Valley homelands to the Mississippi River in the 1550s .

Occurrences within each of these periods in the Quapaws' history, however remote these events and occurrences may seem, have a direct bearing on their destiny including, for a brief time, becoming some of the richest people in the world. The history presented in Part I and the Addendum prepares the foundation for the remainder of the book and the observations and conclusions presented in the Epilogue. However, the casual reader more interested in the mining era and its aftermath may move directly to Part II.

THE QUAPAW—THE DOWNSTREAM PEOPLE

"And the land was not able to bear them, that they might dwell together..."[7]

The attacks by the Iroquois wolf packs and the resulting westward movement of the Dhegiha Sioux from the Ohio Valley
did not occur all at once. But eventually the member tribes of the
Dhegiha Sioux reached the Mississippi just after the mid 1600s.
Tribal traditions indicate that when the tribes reached the great river,
the Osages crossed in skin-covered boats. A heavy mist arose and
prevented the remaining tribes from following. The Omahas went
northward and crossed the Mississippi near the location of present-day Des Moines, Iowa. Further separation among these tribes
occurred as they continued their westward movement into western
Missouri, eastern Kansas, western Iowa, and northeastern Nebraska.
The Quapaws went downstream and eventually settled in four villages on both sides of the Mississippi near the mouth of the Arkansas River, far removed from the general location of the other four
Dhegiha Sioux tribes. The Quapaws forcefully displaced the Tunica
and Illinois Indians and occupied the abandoned villages. The hated
Chickasaws resided on the east side of the Mississippi and would
be involved in many conflicts with the Quapaws in the decades to
come.[8] The Osage would eventually settle along the Osage River
in Western Missouri and be the closest of the Dhegiha tribes to
the Quapaws. Osage hunting parties would range southward into
Arkansas[9] resulting in periodic conflicts with the Quapaws.

Another Quapaw legend recounts a similar story. When the
wide river was reached on their westward journey, the people made
a rope of grapevines. They fastened one end on the eastern bank
and the other end was taken by strong swimmers and carried to the
western bank. The people began to cross the river by clinging to
the grapevine. When about one-half of the people had reached the
western side, the grapevine broke, leaving the Omahas and Iowas

MIGRATION OF THE DHEGIHA SIOUX FROM THE OHIO VALLEY.

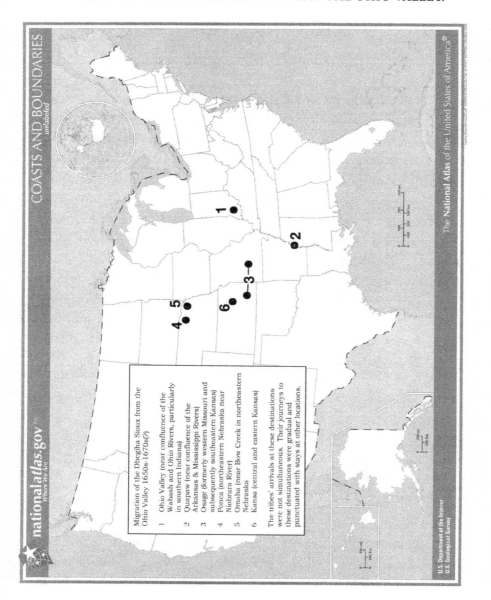

Figure 1

behind. Left also were the Quapaws. The crossing had been made on a foggy morning and those left behind, believing that their kinsmen who had crossed had turned downward on the western side, turned downward on the eastern side. Thus the two groups were separated. The Quapaws continued downstream until eventually they reached their Arkansas location.[10]

The Quapaw name is said to have been derived from Ug'akhpa, meaning "Downstream People." Thomas Nuttall in his journey into Arkansas Territory in 1819 called them O-guah-pa. Various spellings were given the name over the years, but the guttural sound of their own pronunciation of Ogupas probably evolved into use of the word Quapaw. The Illinois tribe called them Alkansa or Arkansa, a term rarely if ever used by the Quapaws according to Nuttall. Nuttall also stated that the Quapaws were sometimes called the Osark. Nuttall surmised that this name came from French residents of the country who applied the name of the river to the people—Riviere des Arks or d'Asark from whence Osark was derived.[11]

Quapaw life revolved around their villages. Kappa was located twenty-one miles above the Arkansas River on the west bank of the Mississippi. Tongigua was eleven miles above the Arkansas but on the east bank. Tourima was located on the north side of the Arkansas near its mouth, and Osotouy was sixteen miles to the northwest, also on the Arkansas. Each village contained at least one large structure that could hold several hundred tribe members and was dedicated to public gatherings. Family life was centered in groups of multiple family houses built of long poles forced into the ground and arched at the top. The framework of polls was covered with cypress bark and cane mats. The shelter contained a fireplace for each individual family unit and raised sleeping platforms around the perimeter. Village meetings were held in a large bark-covered long house that held as many as 200 people. A plaza with an elevated structure at one end was located at the center of each village. The chief and other village leaders would observe village festivities from this platform.[12]

In 1682, just nine years after their first encounter with the French,

QUAPAW VILLAGES—1682

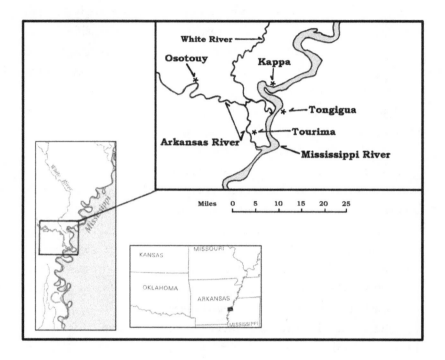

QUAPAW VILLAGES – 1682

Near the confluence of the Mississippi and Arkansas Rivers

Figure 2

a Jesuit missionary estimated the population of the Quapaws to be between 15,000 and 20,000, but this number was believed to be an exaggeration. Corn had become the main staple of the Quapaw diet and required fields containing thousands of acres. If the population was remotely close to the missionary's estimate, it is easy to see that such large fields of corn would be needed to sustain the Quapaw. The Quapaws also raised gourds, pumpkins, sunflowers, beans, squash, and a variety of fruits as well as domesticated chickens, turkeys, and bustards.[13] One of the first Frenchmen that encountered the Quapaws reported seeing fields as large as four or five square miles containing corn, pumpkins, sunflowers, beans, peaches, plums, persimmons, watermelons, mulberries, grapes, and other fruits and vegetables.[14]

When encountering the Quapaw men for the first time, sixteenth and seventeenth century Frenchmen would describe the Indians as tall, large, handsome, well made, and well proportioned. In addition to their beauty, the Quapaws were perceived as honest, generous, cheerful, lively, and polite in contrast to other tribes encountered by the Europeans.[15] Clothing was of little importance to the Quapaws with men only wearing buffalo robes during the winter months. According to their French observers, the majority of the year the men were "stark naked" with the women being "half naked."[16] What the Quapaws lacked in clothing they substituted for with elaborate ornamentation, hairstyles, and body paint. The men plucked their body hair and shaved their heads except for a scalp lock, a narrow center tuft on top of their heads. Most wore beads in their nose and ears with some attaching a horse mane to their hips. Bodies covered with red and black paint were adorned with feathers, skins, and animal horns. Married women wore their hair in a loose, unadorned lock while unmarried women braided their hair in two plaits wound into buns at their ears.[17]

Dance ceremonies were an important part of the tribe's social and political life and were often performed when the Quapaws were dressed in their full regalia. There were dances for almost every occasion including birth, marriage, death, war, peace, hunting, harvest (Green Corn Dance), and fertility. Baird refers to a sensuality dance held secretly by the light of a large fire at night. "Both men and women danced completely nude, coordinating their poses and gestures with songs that expressed their sexual desires."[18] Jean-Bernard Bossu was a Frenchman who arrived at the Quapaw villages in 1751 and made two subsequent visits with the last in 1770–1771.[19] Bossu described his observations of this last dance in a letter to a friend.

> When I arrived among the Arkansas, the young (Quapaw) warriors welcomed me with the calumet dance. I should tell you, sire, that these people dance for many different reasons: there are dances dealing with religion, medicine, joy, ritual, war, peace, marriage, death, play, hunting, and lewdness. This last dance has been abolished since our arrival in America.[20]

The family unit was the basis for the Quapaws' social organization. Marriage was generally monogamous and between non-relatives. The unceremonious, or "blanket," marriage was the most common form of union and consisted of nothing more than a man and woman agreeing to live together. The seeming ease with which the Quapaws had separated themselves from relationships with their other Dhegiha Sioux kinsmen was also reflected in the marriage relationship. Divorce, or "quittings," was frequent and easy. The husband could virtually walk away from the relationship if he deemed the wife to be loathsome to his family. Additionally, the father of the wife could recall his daughter if he perceived the son-in-law was mistreating his wife.[21]

George Sabo III presents remarkable insights into Quapaw society as it began interacting with the Europeans during the Quapaws sojourn at the mouth of the Arkansas River. Sabo states that the Dhegiha Sioux kinship system, both in marriage and through alliances, was an important element in the provision of unity through reciprocal obligations and rights. Such incorporations and alliances were important to "…bring order and predictability to situations that were inherently unpredictable and, as a result, potentially uncontrollable and dangerous."[22] This order among individuals and groups also gave structure to economic, political, and other institutions.[23] Within the Quapaw society of the time, clan organization required that individuals marry outside of their clan, which would strengthen the tribe by establishing ties with other clans. Alliances with outsiders (non-kinsmen) were also used to create order through the mutual acceptance of rights and obligations. The ceremonial smoking of a peace pipe was often the symbol of the formation of an alliance and therefore acceptance of certain mutual rights and obligations. To the Quapaw, the Wahkonda (Wah-kon-tah) was the invisible life force that brought order, possible only through such a supernatural power. The smoke ascending from the peace pipe was believed to be the smoker's breath conveying his desires to Wahkonda. As Sabo states, "The Sacred Pipe thus could serve effectively as an arbiter of peaceful contact and interaction between unrelated groups because its use imposed the inviolable sanction of supernatural forces upon statutes and relationships negotiated between groups."[24]

Of all the European groups during the 200 years following the first contact of Europeans with the Quapaw, the French were their longest and closest allies. But, Sabo would call the French "…inconsistent kin, individuals incapable or unwilling to live up to the reciprocal obligations the Quapaws expected of them on the basis of their alliance."[25] However, that same inconsistency appears evident in the Quapaws themselves in regards to their casualness with which they allowed relationships to end both on family and tribal levels.

The hereditary chief made decisions affecting the life of a Quapaw town. The chiefs of the four towns would discuss and make decisions in matters affecting the entire tribe. Respected counselors called elders gave counsel to the chief. Additionally, a group of young men that had demonstrated wartime prowess were selected to serve the chief. Below the elders and selected young men in the village hierarchy came the other men followed by the women and children. Slaves were not considered part of the Quapaw hierarchy. These Indian captives resulted from battles with neighboring tribes, some of whom the Quapaws had expelled upon their arrival at their new Arkansas home. Some slaves were captured in encounters with other tribes such as the warlike Chickasaws east of the Mississippi. The Quapaw women determined if a captive was to live as a slave, perhaps to replace a husband or son lost in battle, or to be slowly burned to death over a fire.[26] However, the description of the afore-mentioned Quapaw chain of command is not as precise nor those at the top of the chain as powerful and authoritative as it may appear. Generally, the Quapaws operated on the basis of consensus at council meetings. A chief's power and influence would vary depending on his reputation, wisdom, and general leadership capabilities.[27] A chief could not demand but only coax his people into following some course of action that he felt was important for the tribe.[28]

Traditions of the Dhegiha Siouans indicate a belief in an origin in the Sky World. As late as the nineteenth century, there were bits and pieces of customs and traditions that indicated worship of the sun, moon, and stars.[29] The Quapaws believed an individual would exist after death but be judged and receive a life of joy or be subjected to a perpetual life of torment based on that judgment. As previously stated, the Quapaws believed the Wahkonda was the central force of the universe and the life power inhabiting nature and mankind. Because of this association of nature and mankind through the Wahkonda, the Quapaws believed they were related to many natural phenomena, both animate and inanimate. Wapinan (holy men) were responsible for consulting Quapaw deities.[30]

In 1673, soon after their arrival at the mouth of the Arkansas, the Quapaws had their first encounter with Europeans. Two Frenchmen had led an expedition down the Mississippi with the purpose of expanding France's fur trade on the continent and spreading Christianity. Jacques Marquette was a frail Jesuit priest with an interest in Indians, master of six Indian languages, and described as unusually sweet and gentle. Louis Joliet was a twenty-eight-year-old former Jesuit student turned fur trapper. His giant physique was tempered by a tactful and sensible nature.[31] The two men plus five others departed on May 17, 1673, from St. Ignace at Michilimakinac, a missionary outpost on Lake Huron.[32] By way of Lake Michigan, then overland to the Wisconsin River, the party arrived at the Mississippi on June 17. Descending the Mississippi, the explorers eventually approached a village near the mouth of the Arkansas occupied by the Mitchigamea, an Algonquin fragment that would later be driven out by the Quapaws. Ironically, remnants of both tribes would be settled 150 years later in Indian Territory. The area was located in the far northeastern corner of what would eventually become a part of the new State of Oklahoma in 1907. After a night with the friendly Mitchigamea, the seven men traveling in their two birch-bark canoes continued down the Mississippi and soon arrived at the shore of the northern most Quapaw village changing the lives of the Quapaws forever.[33]

News of the arrival of strange white-skinned men had already reached Kappa, the northern-most village of the Quapaw on the western bank of the Mississippi. Two canoes of Indians met Marquette's party a mile or two north of their village. The leader signaled their peaceful intentions by holding a calumet high above his head. The visitors were given food and tobacco and asked to follow the Indians' canoes to their village downstream where the men paddled ashore on that fateful July day in 1673. An Algonquin captive informed the Marquette and Joliet that they had arrived among the "Akamsea." Marquette and Joliet told the Indians of their quest and gave an explanation of God and their Christian faith. The men were told that the river pathway they thought led to the Pacific Ocean actually emptied into the Gulf of Mexico. The Quapaw also warned the Frenchmen that that the tribes downriver were warlike. After a feast including dog meat with their Quapaw hosts, the disappointed men rested a day and began their return journey up river to Lake Michigan on July 17.[34] Another account states that the Quapaws would first threaten the Frenchmen and only refrain from attack because of the peaceful overtures of Marquette as he held the calumet high over his head.[35] There was a group of Quapaws who wanted to kill and rob the Frenchmen as they slept, but the chief's opposition to the warriors and vigil before the lodge where the Frenchmen slept insured their safety.[36]

Some historians have challenged the theory that the Quapaws originated in the Ohio Valley. They suggest that the Quapaws' first encounter with the white man occurred when Hernando De Soto, Governor of Cuba, discovered the Mississippi River in June 1541 and ascended to the mouth of the Arkansas River. De Soto encountered the Pacahas whom some would claim to be the same people as the Quapaws found by Marquette occupying the same region over 125 years later.[37] Although a lively and ongoing debate continues, historians, archaeologists, and most ethnologists have rejected the once-popular assumption that the Pacahas met by De Soto were the Quapaws. The Pacahas are believed to be a branch of the Tunica, a tribe displaced by the Quapaw after their arrival at the mouth of the Arkansas.[38]

After their encounter with Marquette in 1673, the Quapaws' next contact with the Europeans occurred nine years later. Another Frenchman, Rene'-Robert Cavelier, Sieur de la Salle, would descend the Mississippi in hopes that it flowed to the Vermilion Sea (Pacific Ocean) and thus to China. A second objective was to determine the nearness to Canada of the kingdoms of the Quaio and Quivira where gold mines were believed to exist.[39] La Salle's expedition reached the Mississippi by way of the Illinois River. Hearing war cries and drumbeats as he approached the northern most Quapaw town of Kappa on March 13, 1682, he and his party quickly retreated to the eastern bank and prepared to defend themselves. Henri de Tonti, La Salle's lieutenant, returned to the riverbank and hailed the approaching Indians already at midstream. Determining that the white men were not bent on war after visits by emissaries to each camp, La Salle and his men were soon invited to Kappa where the hospitable Quapaws supplied wood, food, and shelter. The Quapaws' hospitality was fueled by their desire to obtain French weapons with which to subdue and dominate their enemies. Tonti estimated the Quapaw warriors at no more than 1,500 men.[40] Adding the other men, women, children, and slaves, the total Quapaw population would likely have been between 5,000 and 10,000. This was significantly less than a previous Quapaw population estimate of between 15,000 and 20,000.

On March 17, La Salle continued his journey down the Mississippi. He arrived at the Gulf of Mexico and ceremoniously claimed Louisiana in the name of Louis XIV on April 9, 1682.[41] La Salle hoped to exploit the commercial potential of the region (the fur trade of interior North America) and establish a base from which to challenge other colonial powers. To do so he would need the help of friendly tribes such as the Quapaw to establish a series of strategically located posts. La Salle hurried back up the Mississippi to Canada and returned to France to present his plan to the king. Louis XIV was enthusiastic about La Salle's vision and outfitted the explorer with four ships and several hundred troops and colonists. The vessels departed on July 24, 1684, over two years after their initial contact with the Quapaws. The fleet did not turn north at the Mississippi but sailed on to the coast of Texas at Matagorda Bay. Following a shipwreck and mutiny, La Salle led a small group of his men as he wandered around Texas until late 1686 when a mutinous faction of his group killed him.[42]

Henri de Tonti and a group of twenty-five Frenchmen and nine Indians left their Illinois post in February 1686 to search for the missing La Salle expedition. They traveled down the Mississippi to the Gulf, found no trace of settlement, and returned upriver to the Quapaw villages. Tonti consented to leave Jean Couture and five others to establish a post (later known as the Arkansas Post) to wait for the missing expedition and to begin commerce with the Quapaws. Six men from La Salle's group—including his brother, Abbe Jean, and nephew, Henri Joutel—pushed on following La Salle's death and arrived at the Quapaw village of Osotouy in July 1687.

They were met by Jean Couture and the other Frenchmen who had waited at the Quapaw villages since the previous summer for the La Salle expedition's arrival. The men were greeted by Quapaws with the same hospitality as previously shown to La Salle, but the Quapaws were not told that La Salle was dead and that his expedition had ended in disaster. With the Quapaws help, the men continued up the Mississippi. Henri Joutel did not tell Tonti of La Salle's death, and Tonti only learned of it from Jean Couture in April 1688. Tonti was concerned about the remainder of La Salle's party in Texas and proceeded down the Mississippi with a small search party. They arrived at the Quapaw villages in January 1689. The men were again received with great joy by the Quapaw. Tonti's party continued to the southwest but learned from the Caddo that none of La Salle's party had survived. Tonti returned to the Quapaw villages in late July, remained for two weeks while recovering from a fever, and then returned to his headquarters on the Illinois.[43]

For the next ten years, Tonti continued to send his men to the Downstream People to exchange trinkets and a small amount of weapons for the Quapaw pelts. However, for the Quapaws, the unfortunate byproduct of their desire for trade and weapons was devastation by smallpox. By 1699, only 300 warriors remained according to reports of three Seminary of Quebec missionaries who were traveling through the lower Mississippi Valley. By late 1700 the three villages of Kappa, Tourima, and Tongigua had combined at a location on the west bank of the Mississippi just below where Marquette and La Salle had first encountered the Quapaws.[44]

In February 1700 a group of traders sent by British Carolina Governor Joseph Blake delivered presents to the Quapaws hoping for commercial and political alliances. The British desired slaves that could be sold along the Atlantic coast. The Quapaws readily complied by crossing the Mississippi and capturing a number of Chakchiumas, a Muskogean-speaking people that later became part of the Chickasaws. The English continued to court the Quapaws and other Indian people of the area. But through the efforts and leadership of Jean Baptiste le Moyne de Beinville, France had once again established its dominance over the Mississippi Valley by 1713. In the summer of 1721, Second Lieutenant de la Boulaye and thirteen soldiers began residing with the Quapaws and built a military post adjacent the their village on the Arkansas River. Soon eighty German colonists established a settlement on 24,000 acres next to the tribal villages. By 1727, there were no more than 1,200 Quapaws remaining along with a few colonists and no soldiers.[45] The French-Quapaw friendship grew as the popular trading post provided an arena for trading furs and salt. The Frenchmen that stayed began to take the Quapaw women for wives.[46]

The Quapaws continued to encourage the missionary priests to stay among them, and in 1701 Father Nicholas Foucault began a work among the Quapaws. However, members of the Koras tribe killed Father Foucault in 1702. It would be twenty-five years before Father du Poisson began another mission work among the Quapaws in 1727. As with Father Foucault, Father Paul du Poisson would be killed, this time by the Natchez and their allies in November 1729. On a trip to New Orleans, du Poisson had stopped at Fort Rosalie when the Indian attack occurred that killed him and 250 others. Father Avond would continue the mission efforts ten years later. One other missionary priest, Father Carette, would serve the Quapaws prior to France's cession of its territory east of the Mississippi to the British in 1763.[47]

Conflict between the British and French in the Mississippi Valley was a microcosm of a much larger conflict that had been happening since the first explorers and colonists stepped onto the shores of the North American continent. Sporadic raids and retaliation between the British and French and their respective Indian allies along the Mississippi including the Quapaws would continue during the decades of the 1730s and 1740s as a prelude to the French and Indian War. The decline of the Quapaws both numerically and as a viable tribal society began in the Ohio Valley with their expulsion three quarters of a century earlier. Although far removed in time and distance from the conflicts of their ancestors along the Ohio Valley, actions taken by both the British and French in the region would once again portend ominous consequences for the Downstream People. Traders from Pennsylvania rapidly expanded their outposts westward into the Ohio area between 1744 and 1754. Aroused by these trailblazing activities of the British, the French attacked three trading posts, killed their occupants, and established their own forts during 1752 and 1753. Virginia Governor Robert Dinwiddie dispatched twenty-one-year-old George Washington to the forts to protest the French actions and determine their intentions. He returned to Dinwiddie in January 1754 to report that the French intended to occupy the entire Ohio. Hostilities began on April 3, 1754, when the French seized the fort being built by the British at the confluence of the Allegany and Monongahela forming the Ohio River (present-day Pittsburgh).[48] Seven years of fighting ended with the surrender of Canada by the French. To compensate Spain for her involvement, France ceded all of its territory west of the Mississippi including the Isle of Orleans by the secret Treaty of San Ildefonso on November 3, 1762. With the Treaty of Paris signed on February 10, 1763, France ceded all Canada, Acadia, Cape Breton, the islands of the St. Lawrence, and all of its lands east of the Mississippi to Great Britain. Except for a couple of islands in the Caribbean, the French vision of being a major power in the Western Hemisphere ended.[49]

Ninety years had elapsed between the Quapaws' first meeting with Marquette on the banks of the Mississippi in July of 1673 and the summer of 1763 when they realized the implications of France's defeat and the cession to Spain of lands west of the Mississippi. The Quapaw chiefs traveled to New Orleans and pleaded with the French governor, who was sympathetic to their plight but powerless to help. Perhaps it is not too difficult to imagine that as the chiefs paddled their canoes northward toward their villages, they reflected on their history since meeting the first white man ninety years earlier. With an estimated 6,000 to 15,000 tribal members in

1682, the tribe was now less than 700 in number, of which only 160 were warriors. Although some losses occurred from their wars with other tribes, both for themselves and on behalf of their French allies, most of the deaths resulted from three epidemics in occurring in 1698, 1747, and 1751. The 1698 epidemic reduced the tribal population by two-thirds or more. The European presence was the likely source for most of the diseases. The desire for weapons and trade with the white man tended to pull the Quapaw from their agricultural roots and require more time spent in the hunt for hides and away from the tribal villages. The Quapaws' cultural integrity was also adversely impacted by intermarriage of Quapaw women with the white traders and colonists and by the introduction of alcohol, which was particularly devastating to the Indian.[50]

Studies of historical texts written in the last two-thirds of the twentieth century reveal an almost universal condemnation of the Europeans' arrival and expansion into the North American continent. However, when one strips away the current zeal for interpretation of history based on a very narrow political and sociological agenda, the understanding of the era is placed in a perspective that has been well understood and accepted for generations. To be sure, many of the European influences proved disastrous for the resident populations. However, this has been the history of mankind from its beginnings. The North American continent could not have remained isolated forever from the either the good or corrupting influences of other cultures. Although the Europeans and Africans commercialized the practice of slavery, the Indians had practiced a far harsher form of slavery and torture for millennia. The Indians' physiological susceptibility to the diseases brought from foreign shores and the acute effects of alcohol were not tools of oppression by the white man but byproducts of a meshing of cultures. Although these weaknesses were used by unscrupulous individuals and groups to achieve immediate gains, it was not a purposeful and ongoing means to conquer the native peoples.

A case in point involves the coastal Indians of southeastern New

England called the Wampanoag, a loose coalition of several dozen of villages. These and other coastal Indian societies had traded with the English that arrived by ship for at least a hundred years prior to the arrival of a small band of dirty, malnourished mariners that struggled ashore on a cold December day in 1620. The Wampanoag and other Indians along the New England coast had traded with the English that arrived by ships but, when necessary, forcefully limited the foreigners' stays ashore and discouraged inland incursions. At the same time, the coastal Indians did not permit the interior tribes to directly trade with the English, which allowed the coastal Indians to leverage their geographic position and thereby increase their trading profits and wealth. Massasoit was the military and political leader of the Wampanoag. On March 17, 1621, Massasoit and two interpreters, Squanto (Tisquantum) and Samoset, approached the mariners who had waded ashore a little over three months before. By the time of Massasiot's initial contact, half of those that arrived were now buried. This colony called Plymouth, occupying a deserted Indian village, was inhibited by a people that would become known as the Pilgrims of the history books. Massasoit offered something new to this little band—to stay for an unlimited time. Many Wampanoag villages had been decimated by disease five years earlier. To a worried Massasiot, the change of policy with regard to limiting the foreigners' stays was a bold move to counter the dangers posed by the Narragansett alliance to the west whose numbers and strength now were far superior to the Wampanoag alliance.[51] In return for the Wampanoag's concession, the Pilgrims agreed to form an alliance with the Wampanoag. The Wampanaog-Pilgrim alliance was successful in that it held off the Narragansett. However, unforeseen by Massasoit, the Pilgrims' permanent settlement would act as the toehold for the coming English migration to New England. By the end of the nineteenth century, this toehold would erase all but a faint shadow of the Indian cultures that abounded hundreds of years before. Writers of the colonial era would attribute the depopulation of the Indian world by disease as the "will of God." Later writers would say that the Indians' demise was a result of the superiority of European weapons and other technologies. To quote Charles C. Mann, "Whether the cause was the Pilgrim God, Pilgrim guns, or Pilgrim greed, native losses were foreordained; Indians could not have stopped colonization, in this view, and they hardly tried."[52] However, like Massasoit, most Indians tried to control their own destinies but fell victim to the law of unintended consequences just as they fell to diseases for which their immune systems had no resistance. And, it must be remembered that disease and plague present on the American continent proved fatal to the Indians and the Europeans alike.

This collision of cultures had been occurring for thousands of years as one tribe or group migrated from one area to another because of drought, flood, famine, danger, invading armies, disease, and countless other reasons. Before the white man arrived on the scene, the Iroquois displaced the Quapaws in the Ohio Valley, who then fled to the mouth of the Arkansas on the Mississippi and who, in turn, displaced the Tunica and Illinois Indians. The advance of civilization is never clean, neat, and fair to all concerned. Since the beginning of man's presence on the planet, many cultures have disappeared through destruction or absorption into another culture. But many cultures have adapted and flourished as encounters with other cultures occurred. Even a half-century before the French and Indian War of the 1750s, the native cultures of the woodlands were on the wane. The disease devastated Indians faced destruction because it could not put its ancient feuds behind them and present a united front to stop the expansion of the Europeans on the continent.[53] But it was the white man's culture that put an end to the constant wars among the various tribes. It also ended the practice of cannibalism, savage torture of captives, and, belatedly, the evil of slavery. The American Indian avoided extinction because of their ability to adapt to change and adjust their lifestyle. Such change, adjustment, and accommodation do not necessarily signal the decline or end of a culture. The pre-Columbian Indian cultures previously discussed reflect a continual adaptation to changes either readily accepted or forced upon them. Even before physical contact with all but a few Europeans, the diseases brought by the first European explorers had spread throughout the continent and caused significant reductions of Indian populations and corresponding social changes. To insist that a people remain unchanged, always doing things the same as their forefathers, is to deny reality in a constantly changing world. Social change will not in itself result in the erosion of ethnicity.[54]

The Quapaws were much more adaptable than many tribes in accommodating the changes in and challenges to their culture. Morris S. Arnold in *Rumble of a Distant Drum* draws insightful conclusions from his exhaustive study of the Quapaw Arkansas society from their first encounter with the white man in 1673 to the end of the French-Spanish influence in the region in 1804. Arnold's research reveals a picture of accommodation and compromise that allowed the Quapaws and their European visitors to co-exist more or less peacefully for five or six generations of their contact. Arnold called this accommodation and compromise by both the Quapaws and Europeans a "dynamic kind of symbiosis." By this Arnold meant that in the Arkansas society of the time, the Quapaws were often able to exercise their strong cultural will and prevail. Likewise and just as often, the power and determination of the European would command the day. Sometimes the contest of wills was such that a third way, a compromise, was necessary in their efforts to co-exist. In all of this give and take of competing cultures, Arnold believes that "...the Quapaw were rational economic actors and their choices therefore inevitably and accurately reflected what they perceived their self-interest required at the time. By self-interest, I [Arnold] do not meant selfishness..."[55]

The unfortunate loss of life through the smallpox epidemic of 1698 was a pivotal event as the tribe lost two-thirds of its population and the critical mass necessary to continue as an important power in the long-term conflicts with other tribes and the European powers present on the continent. Also, the stabilizing work of the priests arrived too late to ward off many of the corrupting elements found among the traders, trappers, and colonists. From their first contact in 1673 until 1739, only two priests ministered to and taught the Quapaws for a total of four years. Other tribes had killed these priests before their work could have better prepared the Quapaws for interaction with the white man's ways.[56] Two other priests arrived and ministered to the Quapaws between 1739 and the French defeat in 1763. However, the damage to the Quapaws had already been done.

Just as the defeat of the French and subsequent ceding of the all lands east of the Mississippi to the British in 1763 would have a significant impact on the Quapaws, so too were the actions of a small band of colonists some thirteen years later and 300 miles to the east of the place where the French and Indian War had begun. On June 7, 1776, Richard Henry Lee stood to his feet in the State House in Philadelphia and spoke these words:

> Resolved [Lee began]:...That these United Colonies are, and of a right ought to be, free and independent states, that they are absolved from all allegiance to the British Crown, and that all political connection between them and the state of Great Britain is, and ought to be, totally dissolved.[57]

Thomas Jefferson was delegated the task of drafting the declaration. Debate continued throughout the excessively hot summer. At eleven o'clock on Thursday morning, July 4, the debate was closed and the vote taken. Twelve of the colonies voted in favor of Declaration of Independence as amended. New York's delegation abstained. The American Revolution had officially begun, but the actual signing of the declaration would not occur until August 2 after a suitable copy had been prepared.[58] That famous portion of the second paragraph of the Declaration of Independence would hold little meaning for the Quapaws for many decades to come:

> We hold these truths to be self-evident, that all men are created equal, that they are endowed by their Creator with certain

unalienable rights, that among these are life, liberty, and the pursuit of happiness.[59]

During the fifty-five years following France's cession to Spain in 1763 of the lands west of the Mississippi to the tribe's signing of its first treaty with the American government in 1818, the Quapaws' bravery and their strategic location on the Mississippi at the mouth of the Arkansas continued to command respect from France, Spain, Great Britain, and the newly independent United States.[60]

As Spain replaced France as ruler of the Louisiana territories in 1763, Spanish concern shifted to the British occupation of the regions east of the Mississippi. English officials attempted to win over the Quapaws' loyalty—and more importantly to discourage interference with English convoys moving north on the Mississippi—by showering the Quapaws with gifts. Their cooperation was sought through both Indian emissaries and direct negotiations. Spain countered the English efforts with her gifts to the Quapaws. The Spanish were less adept at negotiations and the ceremonial trappings of such meetings so important to the Indian chiefs. The British trade terms were much more favorable to the Quapaw than those of the Spanish. However, the Spanish drove the English merchants from the area and destroyed their trading posts, and Quapaws once again were forced back into the Spanish fold.[61]

Spain's support of the rebellious American colonists was an effort to slow British expansion in North America and limit her influence among the Indians. However, the Quapaw had little interest or participation in the American Revolution. By the terms of the Treaty of Paris in 1783, Great Britain relinquished all of its land east of the Mississippi to the victorious new nation. However, Spain's hopes of enhancing its domination of the Mississippi Valley and lands to the west through the United States' victory over the British were soon dashed as the land and trade-hungry former colonists looked to the vast stretches of land west of the Mississippi. In June 1784, to gain the full and unified support of her Indian allies, Spain negotiated a permanent peace between the Quapaws and their sworn enemies for generations, the Chickasaws. With the generous gifts of the Spanish and freedom from threats of war from the Chickasaws, the Quapaws were allowed to enjoy the benefits of trade with the Spanish. Such trade included alcohol, and the quantities supplied by the Spanish became so great that the Quapaw chiefs pled with the Spanish governor in New Orleans to prohibit its importation into Arkansas. Following several Quapaw deaths from drunken brawls, one murder, and the widespread dissipation of tribal members, the Spanish governor ordered an end in June 1687 of the liquor traffic reaching the Quapaw at Fort Carlos, a trading post near the Quapaw villages.[62]

In 1793, French officials again arrived in the United States hoping to involve the young country in its European war with Spain and to gain support for attacking Spanish interests in Florida and Louisiana. Facing continued pressures from increasing numbers of Americans in the Mississippi Valley and from threats of an invasion from the east, Spain ceded its lands west of the Mississippi back to the French at the century's end.[63]

By 1800, it had been 118 years since the French explorer La Salle stood at the mouth of the Mississippi in 1682, planted a cross in the sand, and read a declaration taking possession for France of the entire Mississippi Valley in the name of Louis XIV. New Orleans had been established in 1718, becoming the important and prosperous city sitting beside the mouth of a river that drained half of a continent. The Quapaws had only come to this vast land less than two or three decades earlier when La Salle issued his proclamation before a small group of Indians listening to this strange foreigner. With the Treaty of Fountainebleau in 1762, the territory had passed from France to Charles III of Spain. Thirty-eight years would pass and Charles IV would negotiate the secret Treaty of San Ildefonso with Napoleon Bonaparte on October 1, 1800, whereby France once again took possession of the territory.[64] Rumors of the treaty reached President Thomas Jefferson. Jefferson feared the threat of a nearby aggressive imperial power and that the French would end the Americans' right to deposit cargo in New Orleans. He also feared the French

LOUISIANA PURCHASE—1803.

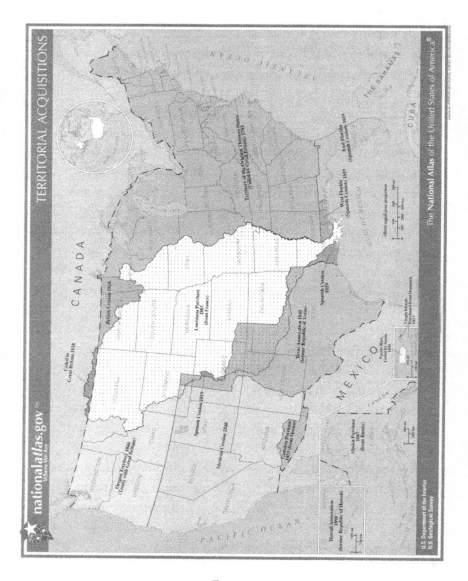

Figure 3

would prohibit passage on the Mississippi that had been secured by the three-year treaty with Spain in 1795. Although the treaty had expired, the Americans continued to enjoy those rights with the Spanish in New Orleans. Closure of the port at New Orleans by the French and the loss of free passage on the river would endanger the entire economy of the United States' western territories. By October 1802, Charles IV signed the official decree transferring the domain to France.[65]

Even before signing of the treaty and its official announcement, France had ordered 5,000 to 7,000 troops stationed in its Caribbean colony of St. Domingue (Haiti) to sail to and occupy New Orleans in March 1802. However, when revolution on the island and an outbreak of yellow fever prevented their departure, Napoleon ordered troops to be sent from the French-controlled Netherlands. The order was given in June of 1802, but it took seven months to assemble the necessary men and ships. For a second time French troops were prevented from sailing to New Orleans, this time by January ice blockages at the port. Meanwhile, Jefferson ordered James Monroe, a former congressman and future president, to travel to Paris with authority to spend up to $9,375,000 to secure New Orleans and parts of Florida. Monroe arrived in Paris on April 12, 1803. Just the day before, Napoleon had decided to sell all of Louisiana. The French finance minister demanded $22.5 million. The Americans offered $8 million. On April 29, the parties agreed to $15 million, or about three cents per acre, for the 830,000 square miles spreading from the Mississippi River to the Rocky Mountains and from the Canadian border to the Gulf of Mexico. The agreement was signed on May 2 but backdated to April 30. The Americans lacked the authority to buy the entire territory and did not have sufficient funds to pay. With contacts at Britain's Baring & Company bank, the French Finance Minister arranged for the young republic to borrow the necessary funds from Baring and a consortium of other banks. The United

States would issue bonds to be repaid over fifteen years at 6% interest. The loan arrangement with the British banks was ironic in that Napoleon would use the funds in his war with the English. Also, the loan was made in spite of Britain's offer of a $100,000 bribe to Napoleon's brother, Joseph, to persuade Napoleon to not sell Louisiana to the Americans.[66]

Apart from the Declaration of Independence and the founding of the United States, perhaps no other event ranks in importance with the acquisition of the Louisiana Purchase in establishing the nation as a continental power and ultimately a world power. With the stroke of a pen, the size of the United States was doubled, and in time the Louisiana Purchase would provide space for all or parts of fifteen states. At the time of the purchase, the non-Indian population of whites, slaves of African origin, and free persons of color in the territory was estimated at approximately 8,000,[67] one person for every one hundred square miles acquired. That would soon change.

The Quapaw—Old Americans or New Americans?

"Lo, then would I wander far off, and remain in the wilderness..."[68]

The new American government's diplomatic efforts with the Quapaws immediately following the Louisiana Purchase appeared to be as ineffective as those of their predecessors, the Spanish. Early American efforts at winning over the Quapaws included the delivery of presents of tobacco and whiskey to the villages in May 1804. This was followed by the stationing of sixteen troops at the Arkansas Post, formerly Fort Carlos but renamed Fort Madison. In 1805 the American government established a non-profit trading post that dispensed cheaper but higher quality merchandise. But the Quapaws wanted more than commerce. The generous presents from France and Spain in bygone years were not forthcoming from the Americans.[69] In 1807, President Jefferson sent John B. Treat to investigate conditions in the newly acquired Louisiana Purchase. Treat was appointed temporary agent of Indian Affairs and assigned to the Arkansas Post among the Quapaws. Treat reported that the Quapaws were friendly, well-respected Indians who hunted, farmed, raised horses, and were known for their bravery in time of battle. As the French and Spanish had done in times past and because the Quapaws expected it, Treat recommended an annual distribution of presents of powder, lead, blankets, and cheap cloth made from woolen rags. A census of the Quapaws in their three villages reflected a population of 195 warriors, 254 women, and 106 children under the age of fourteen.[70]

Not only did the Quapaws worry about their status and relations with the new American government, the tribe must have thought that even nature had turned against them. At 2:30 a.m. in the morning of December 16, 1811, a series of earthquakes of an historic and previously unknown geologic magnitude began along the Mississippi River near the intersection of what are now Missouri, Arkansas, Kentucky, and Tennessee. The epicenter of the first of five earthquakes, each estimated to be 8.6 on the yet to be invented Richter scale, occurred only 150 miles north of the Quapaws' location at the mouth of the Arkansas. Yet the quake was strong enough to cause President James Madison to be shaken out of bed at the White House in Washington, D.C., over 800 miles away. These infamous New Madrid earthquakes were a series of over 2,000 tremors occurring over five months with perceptible aftershocks occurring several years thereafter. The earthquakes were centered near present-day Blytheville, Arkansas, on the Mississippi River and northward along the river to New Madrid, Missouri. Two additional 8.0+ magnitude earthquakes occurred that first day. The second of the three occurred about 8:00 a.m. just north of the present-day Blytheville, and the third occurred at 11:00 a.m., destroying the town of Little Prairie where present-day Caruthersville, Missouri, is located. Earthquakes of 8.0 magnitude are a rarity and generally occur only once each year, sometimes occurring as infrequently as once every three or four years.[71] The last two of the five 8.0+ magnitude earthquakes occurred January 23 and February 12, 1812. This last earthquake would be the largest of the five at an astonishing 8.8 magnitude, one of the largest in the world and second only to an Alaskan earthquake in American history. The quakes caused the collapse of miles of riverbanks, boats, and large areas of the earth to sink, the formation of new lakes, and a change in the course of the Mississippi River. Sometimes the ground would open up and swallow the river as well as towns. Then just as suddenly close with such force as to shoot water as high as the tallest trees. Five towns disappeared completely. The Mississippi River ran backwards for a few hours and created two waterfalls lasting for two or three days, each with a vertical drop of six feet.[72]

Earthquake intensity maps of the 1811–1812 earthquakes rank the level of damage caused by the earthquakes on a scale from one to eleven with eleven being labeled "disastrous" and ten being labeled "devastating." The Quapaws' homes were only fifty miles south of the area of greatest damage and within the area of the next greatest damage level.[73] Thomas Nuttall, an eminent botanist and zoologist, traveled down the Mississippi and up the Arkansas River in late

1818 and 1819. While on the Mississippi between New Madrid and Blytheville, Nuttall noted the fantastic devastation caused by the gigantic earthquakes that had occurred seven year earlier. Because of the frequency of the earth tremors "with two or three oscillations being sometimes felt in a day," people were discouraged from settling in the area. Because of the immense losses of life and property and the general devastation in the region, the government allowed the region's refugees to claim public lands available in Arkansas in amounts similar to those lost in the earthquakes.[74] In addition to encroachment of white settlers with legitimate losses, the government's action became an excuse for many others to exploit the chance to claim Indian lands.[75]

The Quapaws were unhappy that the Americans had not held talks with them as they had with other tribes. As a result, the Quapaws threatened to join their Osage kinsmen in war on the Cherokees immigrating to the Louisiana territory. Such a war would hinder the United States' efforts to encourage the Cherokees to exchange their eastern land for new lands in the west. In addition to the threatened war, the establishment of white settlements in Arkansas and farther south in the Quachita River Valley during and after the war of 1812 presented a second obstacle to the United States' policy of removal of the eastern Indians to lands west of the Mississippi. By October of 1815, the white settlers were complaining of the regular killing of their hogs and cattle and the stealing of their horses, as well as violence against the settlers themselves.[76]

The responsibility for handling the United States' affairs in the west including that of the many Indian nations fell to William Clark of Lewis and Clark expedition fame and now Missouri Territory governor.[77] From Clark's return from the famous expedition in 1806 until the end of his life at sixty-eight on September 1, 1838, he would be the leading representative for the affairs of the United States in the west and would serve six presidents from Jefferson to Van Buren.[78] Although Clark had a liking and respect for individual Indians and a growing compassion for their plight, his duties as the principal Indian agent required him to accomplish the almost impossible task of maintaining peace between the Indians and growing numbers of white settlers while promoting the policies of his government and providing for the Indians' welfare. As to the Indians' welfare, Clark had become convinced that the survival of the Indians would require their separation from the increasing tide of traders and settlers. This separation would necessitate removal of the Indians to Kansas and Oklahoma in order to give them time to develop the skills necessary to adapt to the ways of the white man.[79]

It was this man who held the future of the Quapaws in his hands. During the summer of 1815, William Clark conducted a historic assembly of one of the largest groups of Indians observed by white men just 300 miles north of the Quapaw villages at the mouth of the Arkansas. That same summer, 1815, the conflicts were occurring between the Quapaws and the white settlers in Arkansas and the Quachita River Valley. Gathered at Portage des Sioux where the Missouri empties into the Mississippi, two thousand Indians from over a dozen tribes gathered for the Grand Council near a hundred white tents erected by soldiers as two gunboats moved up and down the Mississippi. By the end of the Grand Council, thirteen tribes signed treaties assuring "perpetual peace and friendship." The departing Indians were showered with whiskey and $30,000 in gifts. However, most of the area settlers called for vengeance rather than reward for the loss of property and life they had experienced at the hands of the Indians.[80]

No sooner than the Grand Council's participants had disbursed, Clark faced the escalating problems with the Quapaw in the fall of 1815. The Quapaw raids on the homesteads of the white settlers and possible war with the newly arrived Cherokees threatened in the eminent failure of the United States' Indian removal policy. Secretary of War William H. Crawford believed the white settlements in Arkansas were contrary to government policy and would impede the government's efforts to induce Indians to exchange their eastern lands for new homes in the west. Decisive action was required or the United States' Indian removal policy would fail. This action came in May 1816 when Clark sent two representatives to order the white settlers to remove themselves from their Arkansas homesteads, inviting representatives from the Quapaw, Cherokee, and Osage tribes to come to St. Louis to negotiate their differences. The first official discussions between the Quapaws and the United States since the purchase of the Louisiana Territory thirteen years before were held in early November 1816. The Osage declined to send a representative. Upon arrival in St. Louis the Quapaws expressed their anger over the lack of gifts from the United States and the intrusion of the white settlers into their territory in Arkansas.[81] One cannot help but believe the Quapaws were insulted by their exclusion from the Grand Council a year earlier and lack of a share of gifts distributed to the attending tribes. Especially galling must have been the fact that the Osages and Omahas, their Dhegiha Sioux kinsmen, had been invited.[82] But the Quapaws came to St. Louis prepared to negotiate. Having observed the newly arrived Cherokees and wanting to adopt their habits and improvements, the Quapaws were interested in getting stock and farming equipment in addition to an annual annuity. The Quapaws offered to cede half their lands to be used by the Cherokees and white settlers. However, Secretary of War Crawford had instructed Governor Clark to not accept any cessions for the Cherokee without agreement with the Eastern Cherokees to cede a like amount of their eastern lands.[83]

Disappointed, the Quapaws and Cherokees returned to their respective homes in late November 1816 with Clark's assurances that the President would be made aware of their situations. However, several events hastened a resolution of their grievances. The white settlers' protests were being heard and given attention to by various governmental officials; whereupon a resolution was adopted that objected to the small group of Indians (the Quapaws) holding title to the vast expanses of land in Arkansas and Oklahoma. Additionally, tensions between the Cherokees and the Osages erupted into open warfare in early 1817. In July 1817, the Eastern Cherokees agreed to cede portions of their eastern lands in support of a home for their western brethren. Finally, a new Secretary of War, John C. Calhoun, took office and brought with him a determination to accomplish the removal of the eastern tribes. In February 1818, given the rapidly changing conditions, Secretary of War Calhoun instructed Governor Clark to ask for a land cession from the Quapaws that would be exchanged for a cession from the eastern Cherokees and provide a homeland for those that chose to move west. Following a visit by Clark's nephew and an invitation to come to St. Louis, Chief Heckaton accepted Clark's invitation and arrived in St. Louis in August. Following extravagant entertainment and many gifts, a treaty was negotiated and signed on August 24, 1818. As important as the gifts and annuity were, the Quapaws believed they had finally received the recognition and status they deserved and had become a valuable ally of the United States.[84]

The treaty provided that the Quapaws would be allowed to hunt on the ceded lands, to received goods and merchandise valued at $4,000, and to receive a yearly stipend of $1,000 in goods and merchandise. In exchange, the Quapaws ceded to the United States any claims to lands on the north and south banks of the Arkansas and east bank of the Mississippi except for a reservation on the south bank of the Arkansas between Little Rock and the Mississippi River. The ceded lands obtained by Clark encompassed all lands in Arkansas except a small portion in the southwest corner south of the Red River, all lands in Oklahoma north of the Red River and south of the Arkansas to its confluence with the South Canadian River; from that point, all Oklahoma lands north of the Red River and south of the South Canadian; and to an uncertain extent, cession of all lands in New Mexico between those two rivers. Additionally, the cession included a relatively small tract in Louisiana stretching from the Red River just south of the Arkansas and Louisiana border east southeastward to the Mississippi River.[85] However, Clark had made an error in his description of the ceded lands. The Quapaws could not legally cede that portion between the Canadian and Red Rivers west of the Louisiana Purchase to the United States as the sources of those rivers originated in land then owned by Mexico and were not part of the Louisiana Purchase. Therefore, the real western boundary of the ceded lands was the western boundary of the lands purchased from the French in 1803.[86]

All totaled, the cessions were estimated to be approximately thirty million acres. The Quapaw reserve was estimated at two million acres.[87] Others have estimated the reserve at one and a half million acres[88] and as small as 1,163,704.75 acres.[89] Whatever the actual size of the reserve, the Treaty of 1818 was ratified by Congress. But when the published edition of the treaty appeared, a discrepancy in the description of the reserve's boundaries eliminated approximately 800 square miles—one-fourth of the original reserve approved by Congress. The original treaty language specified "...up the Saline Fork to a point, from whence a 'due north east' course would strike the Arkansas river at the Little Rock..." but was corrected in the published version to conform to the original map at the top of the treaty that showed a "due north" line to Little Rock. William Clark would certify that the "northeast line" was favorable to the Quapaws. However, the original map drawn at the top of the treaty depicted a "due north" line as opposed to a "due northeast" line. The treaty's language would be changed to conform to the map in spite of the protests by the Quapaws and agreement by the territorial governor that an error had occurred.[90] Whether or not the Quapaws were cheated out of the additional land is open to debate. Certainly, those drafting the treaty made an error, but there are solid arguments to support both sides of the debate as to what the drafters intended. On the one hand, an error in drawing the map would have been more obvious and therefore corrected in the drafting process as opposed to catching an error in the written description of the real estate. Therefore, it is more likely the error would have been made in the description of the direction ("due northeast" as opposed to "due north") rather than in the drawing of the map. Yet, it is hard to argue against the certification of the "due north east" line by Clark, one of the architects of the treaty.

The Quapaws were once considered an important ally, first to the French and then to the Spanish. But the decline of the Indian wars, possession of territories on both sides of the Mississippi by one dominant power (the United States), decline of the Quapaw population through disease, and loss of their strategic location at the confluence of the Arkansas and Mississippi Rivers resulted in a dramatic change of status for the tribe. Now the United States considered the Quapaw not as an ally but more of an obstacle to its policy of Indian removal from eastern lands. Previously, the Quapaws dealt with nations headed by monarchies and dictators. But this new nation was driven by leaders elected by the people, and this people wanted land for their growing nation. The cessions first proposed by the Quapaw to Governor Clark in November 1816 would become a reality with the signing of the treaty on August 24, 1818. The consequences of the Quapaws' loss of status in its dealings with the United States would soon be evident.

Prior to the Quapaw Treaty of 1818, the American government viewed the Quapaws as prospective commercial partners. However, with the formation of the Arkansas Territory in 1819, the voices of white settlers began to be heard in Washington. Unlike the representatives of the monarchial or dictatorial homelands in England, France, and Spain, the wishes and interests of the new settlers became significantly more important to the democratic government in Washington. These interests became evident as the first territorial legislature petitioned the President to reduce the Quapaw reserve to no more that twelve square miles. For a time the national government ignored not only the Quapaws but the growing numbers of white settlers as well.[91] As has been previously mentioned, the 1803 non-Indian population of whites, slaves of African origin, and free persons of color was estimated at approximately 8,000[92] in the entire confines of the Louisiana Purchase acquired from the French. By contrast, the population of non-Indians in 1820 was 14,255 in just the area that would become the state of Arkansas.[93]

Within six months of Chief Heckaton signing the 1818 treaty, Congress passed a bill establishing the Arkansas Territory on February 20, 1819. President James Monroe signed the bill on March 2, and General James Miller was appointed governor of the new Arkansas territory on March 3, 1819. However, Miller did not arrive on his specially outfitted keelboat until December. Loaded with gifts for the various Indian chiefs with whom Miller would meet, the keelboat served as Miller's office and living quarters for several months after his arrival in Arkansas. The gifts included four military coats of which Quapaw Chief Heckaton was to receive one. Robert Crittenden, appointed as the territorial secretary, arrived well ahead of Governor Miller and began organizing the territory to his own political liking. Miller was not alone in his dislike for Crittenden, who was described as aristocratic, aloof, brilliant in politics, and totally lacking in diplomatic skills. With good reason, the Quapaws greatly distrusted Crittenden, who would become a principal force for their removal from Arkansas.[94]

Following the creation of the Arkansas Territory in March 1819, the first legislative session of the Arkansas Territory was convened at the Arkansas Post, on the north side of the Arkansas River across from the northeast corner of the Quapaw reserve created by the 1818 treaty. A new town site called "Little Rock" was established in the spring of 1820 on the Arkansas River at the northwest corner of the Quapaw reserve. In the summer of 1820, Territorial Secretary Crittenden and his political associates acquired the town site. In October the legislature reconvened at the Arkansas Post and shortly thereafter passed a bill making the Little Rock townsite the permanent capitol of the territory, even though half of the townsite, containing the new official seat of territorial government, lay within the Quapaw reserve.[95]

In less than three months after its inception in November 1819, the *Arkansas Gazette* began publishing articles and editorials promoting the removal of the Quapaws in their entirety from the Arkansas Territory. As early as February 5, 1820, the newspaper encouraged the appointment of representatives from Washington to meet with the Quapaws to extinguish the Quapaws' claim to lands south of the Arkansas River. The *Gazette* also featured the platforms of various political candidates who supported the Quapaws' removal.[96] Vern Thompson quotes a May 24, 1824, *Arkansas Gazette* article that makes the newspaper's position very clear.

AREAS OCCUPIED BY THE QUAPAW INDIANS
—1550 TO PRESENT

1. Mouth of the Arkansas River at its confluence with the Mississippi River. The four original villages were established near this location after the Quapaws were driven from the Ohio Valley sometime between the 1650s and the arrival of the French at the Quapaw villages in 1673. The 1818 Cession of approximately thirty million acres left the Quapaws with a reserve of about two million acres.

2. Arkansas Post (north bank of Arkansas River near Quapaw village of Osotouy).

3. Quapaw lands following 1818 Cession. Quapaws moved to Caddo lands in Louisiana in February 1826 following the 1824 Cession. The Quapaws that returned to Arkansas were moved to Indian Territory in September 1834.

4. Quapaws on Caddo lands in northwestern Louisiana in February 1826. By fall of 1826, one-fourth had returned to their Arkansas homelands. By September 1830, almost all would return to their former Arkansas homes. In the spring of 1833, approximately three hundred Quapaws followed Sarasin upon his return to join the few remaining Quapaws on Caddo lands at the Red River in Louisiana. By 1835 the Quapaws in Louisiana left for the last time. Fifty joined those on the reservation in Oklahoma, most joined the Cherokees in northeastern Texas, and a few lived among the Choctaws in what is present-day southern Oklahoma.

5. Quapaw Reservation in northeastern Oklahoma – September 1834 (Established by Treaty of May 11, 1833). Due to a survey error, the Quapaw were required to resettle again in late 1838 to just north of where they were originally settled. The resettlement would occur in late 1839, but approximately one hundred Quapaws had left the reservation and joined the Quapaws living with the Cherokees in Texas.

6. Most of the Quapaws living on Caddo lands in Louisiana joined Chief Bowles' Texas Cherokees in northeastern Texas in 1835.

7. Quapaws living with the Texas Cherokees moved to Choctaw lands just north of the Red River in Indian Territory following their eviction from Texas in 1839.

8. Quapaws moved from Choctaw lands south of the Canadian River to Creek lands just north of the Canadian River in 1842 approximately eight miles west of present day Holdenville.

Figure 4

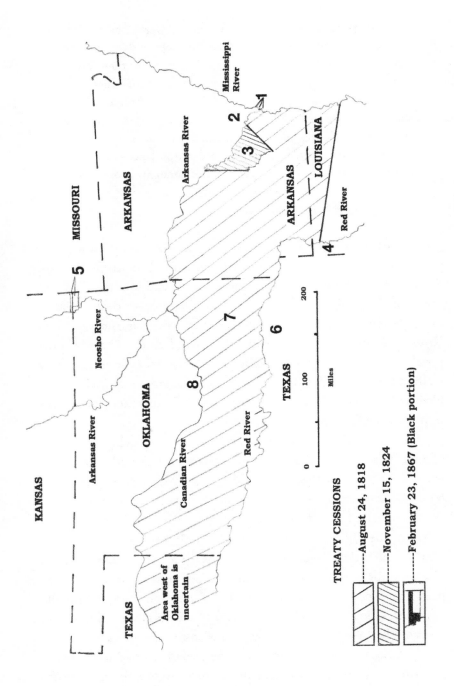

Figure 5

The Quapaws were once a numerous and warlike nation, but, like most other Indians, who imbibe the vices without the virtues of the whites, they retain but a small remnant of their former power, and now number only about 467 souls... These Indians own a vast body of land, lying on the south side of the Arkansas River, commencing immediately below this place, and extending to the Post of Arkansas, comprising several millions of acres, a great part of which is represented to be first-rate cotton land. One of the first wishes of many of our citizens, and of hundreds of others who have visited the Territory with a view of emigrating to it, is for the purchase of the Quapaw lands...[97]

In early May of 1824, Chief Heckaton and seventy-nine Quapaws had traveled to the new territorial capitol in Little Rock to express their refusal to abandon their reserve and join the Caddoes south of the Red River. The Quapaws did agree to cede all lands except a ten-mile wide tract between the Arkansas and Quachita Rivers. Although Crittenden initially agreed to the Quapaw proposal in early May, receipt of subsequent instructions from Secretary Calhoun in June following the action of Congress required a complete cession of the Quapaws' Arkansas lands and their incorporation with the Caddoes south of the Red River.[98]

Henry W. Conway, a candidate for territorial delegate in June of 1822, falsely informed the War Department that the Quapaws were willing to sell their lands for $50,000. One year later, Crittenden—now-acting governor of the Arkansas territory—advised the Secretary of War Calhoun that the cost to the government to be rid of the Quapaws was estimated at only $25,000. On May 26, 1824, Territorial Delegate Conway gained Congress' authorization of $7,000 for negotiations that would lead to a treaty with the Quapaws that would grant cession of their lands and merger with the Caddoes. Conway was allowed to carry back to Arkansas a $7,000 warrant allocated for the negotiations with the Quapaws.[99]

Although working toward the same goal of removal of the Quapaw from Arkansas, Crittenden and Conway were political enemies and began exchanging charges and counter-charges involving misappropriation of a portion of the $7,000 to be used to for negotiations with the Quapaws as well as other financial irregularities. Eventually, Crittenden challenged Conway to a duel in which Conway was killed on the banks of the Mississippi in October 1827.[100]

In November 1824, Chief Heckaton and the Quapaws again met with Crittenden and were shocked to hear that the previous agreement was being withdrawn. In spite of vigorous protests by the Quapaws, they had little choice but to accept the demanded cession of all Arkansas lands and removal to the area occupied by the Caddoes.[101] The meeting was held near the Arkansas Post at the home of Major John Harrington, a respected citizen well-known to the Quapaws, and the treaty was signed on November 15, 1824.[102]

In a November 23, 1824, article, following the consummation of the treaty, the *Gazette* praised Crittenden's efforts and gloated at the almost complete victory in securing the Quapaw lands for white settlement.

> Purchase of the Quapaw Lands—We feel highly gratified in being able at this time to congratulate our fellow-citizens of Arkansas, on the complete accomplishment of his desirable object. It was affected, by Treaty, by Robert Crittenden, Esq., Commissioner on the part of the United States, on the 15 inst. Mr. Newton, Secretary of the Commissioner, has (with the consent of Mr. Crittenden) politely favored us with a copy of the Treaty, which we take much pleasure in laying before our readers today. By this Treaty, it will be seen, that, with the exception of four or five sections, which are reserved for the benefit of particular individuals, the Quapaws have ceded the whole of their lands to the United States, and on terms highly advantageous to the later. No doubts can exist of its receiving the sanction of the President and Senate of the United States, and we hope that the lands thus acquired will speedily be surveyed and brought into market.[103]

The self-dealing of territorial politicians, newspaper promotion, and political pressure from white settlers on their Washington representatives doomed the Quapaws in retaining an Arkansas homeland.

The treaty provided that the Quapaws would vacate their Arkansas homelands, move to the country inhabited by the Caddoes (south of the Red River in Louisiana), and assimilate into that tribe. No provision for an assigned reservation was made. In exchange for their Arkansas reserve, the treaty provided goods valued at $4,000 and an eleven-year annual annuity of $1000 in addition to their perpetual annuity. Additionally, a payment of $500 was made to each of the four chiefs and an assignment of 2,320 acres in Arkansas to twelve persons of Indian descent.[104] In addition to $1,500 removal expenses, $15,372 was allocated for a six-month supply of food.[105] Accompanied by a territorial representative, the Quapaw chiefs traveled to the Caddo lands to scout the prospective new home for the tribe. However, neither the Caddoes nor their Indian agent had received official instructions with regard to the incorporation of the Quapaws into the Caddo lands. Ultimately, agreement was reached with the Caddoes to allow the 455 members of the Quapaw tribe to settle on the land. The tribe now was comprised of 158 men, 123 women, and 174 children.[106]

Over 150 years after their arrival at the mouth of the Arkansas, and in spite of pleas to the new territorial governor George Izard that they be allowed to remain, the Downstream People began the slow trek from their Arkansas homeland to the distant Caddo lands in early January 1826. The first contingent of the tribe arrived on the north bank of the Red River on February 13, 1826. Due to continued resistance by the Caddoes, the Quapaws did not move across the river until March 1. They quickly built three villages approximately thirty miles northwest of present-day Shreveport near the agency compound on the Caddo Prairie. Corn was planted, but flooding of the Red River in May destroyed the crop. A second crop was planted but was again destroyed in June. Subsequently, sixty tribe members died from starvation.[107]

Given the terrain of the area, the spring flooding of the river should not have been surprising to the Quapaw. The location of the Quapaws in Louisiana was less than a dozen miles south of the Arkansas-Louisiana border on the west side of the Red River as it

THE QUAPAWS WITH THE CADDOES—1826-1835.

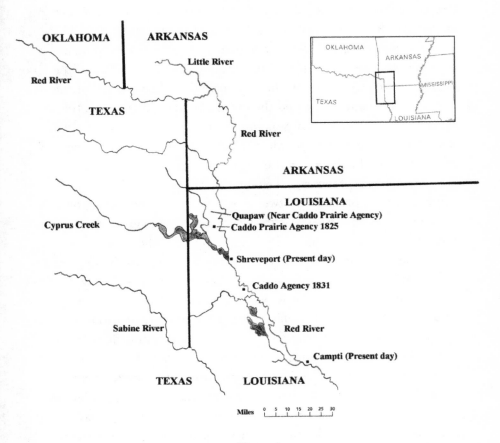

Figure 6

took a southward course beginning a few miles north of the border. For good reason, the Treaty of 1818 made reference to the area as the "Big Raft," a longstanding logjam of monumental size. The southern end of the Red River logjam was located near present-day Campti, Louisiana, approximately a hundred miles south of the location of the Quapaws on the Caddo Prairie. The erosion of the river's banks upstream created the logjam by undercutting the trees along the river. The logjam was intermittent with sections of open water but had extended a hundred miles upstream by the 1830s with its northern end adjacent to the Quapaws' new villages. While the northern end was growing at the rate of a mile per year, the southern end would be approximately twenty-five feet thick. Sections would become silted and form a base for plants and trees to grow. Swamps and lakes would form along its length. By 1833, the Army Corps of Engineers began the multi-year task of breaking up the logjam. Efforts to keep the logjam from reforming continue even today.[108]

In September 1826, a little over six months after the Quapaws' arrival, a half-blood tribal chief named Sarasin led one-fourth of the tribe in an unauthorized return to their ancestral homelands in Arkansas. Sarasin and his group returned to land that had been reserved by him near the Arkansas Post. After gathering for a council on January 28, 1827, Sarasin and his remnant addressed a letter to the President of the United States asking that the Quapaw in Arkansas be placed under the president's protection. Sarasin stated that "our white brothers love us; we have always kept the path clean, a drop of white blood has never stained it."

Sarasin's claims of white friendship were true, proven by an 1825 incident just before the Quapaws' departure for the Red River. The Chickasaws had kidnapped two white children of a trapper, and Sarasin listened as the trapper pleaded for help. Sarasin swam the Arkansas and entered the Chickasaw camp after darkness. He began a war whoop and in the ensuing confusion took the children back across the Arkansas before the Chickasaws discovered that a war party was not present among them. Sarasin became a local hero for the townspeople of Pine Bluff.[109] When the residents of the Arkansas Territory learned of the letter to the president and the plight of the Quapaws, there were cries for action to relieve the tribe's suffering. Some provision was made for both groups of Quapaws. The Arkansas contingent was granted a temporary right to stay in Arkansas and subsisted by building homes in isolated areas. They planted small fields and were employed by white settlers as hunters and cotton pickers.[110]

Governor Izard arrived in Arkansas Territory in May 1825. He had arrived after the Quapaws had signed the November 1824 treaty but before their departure for the Red River. Chief Heckaton highly regarded Izard as a man worthy of confidence. Izard was an old state politician, an ex-Baptist preacher, agreeable in appearance and social habits, and known for keeping his promises to the Indians.[111] After the Quapaws' removal, Izard had personally visited the Quapaws on the Red River. He knew of their desperate situation and would not send the Arkansas Quapaws back to the Red River debacle. John Pope became Arkansas territorial governor in 1829 following George Izard's death. Pope befriended the Quapaws and became known as one of the most humane governors in its history. Pope continually pleaded the cause of the Quapaws. Following the death of Captain George Gray in November 1828, Jeheil Brooks assumed the role of Indian agent for the Caddoes. Gray had been openly hostile to the Quapaws from the very beginning of their stay on the Red River. He did little to alleviate their suffering and starvation while the Quapaws were under his charge. But Brooks, like the new territorial governor in Arkansas, was sympathetic to the Quapaws' plight. He reported their suffering condition to the Secretary of War in August 1830 and provided personal funds for their relief.[112]

Brooks' efforts would prove too late. Small groups of the Louisiana Quapaws began returning to their Arkansas homeland in 1829 and 1830. By September of 1830, only forty remained. By October, only a few Quapaws remained among the Caddoes. In December 1830, Chief Heckaton traveled to Washington to intercede with the federal government on behalf of his people. A request by Heckaton to Secretary of War John Eaton that the Quapaws be assigned tribal lands near their ancestral home was denied. However, Eaton agreed to allow the Quapaws to continue to reside in Arkansas until the land was sold or a new home for the tribe could be found.[113]

Furthermore, most of the promised annuity payments were never paid or paid with such irregularity as to cause severe hardships for the destitute tribe. In January 1832, a new commissioner of Indian Affairs suspended the annuity payments altogether because of the tribe's unauthorized return to Arkansas. However, the new territorial governor John Pope believed the Quapaws were not a threat and had been seriously defrauded. The governor's concern sparked the government to reassess its policies regarding to the Quapaws. But time was never an ally to the Quapaws, as the gears of the white man's government turned slowly. The tribe continued to suffer as eviction from their isolated homes forced them into the surrounding swamps. Newly appointed Indian sub-agent Richard Hannum began pressing for an immediate resolution of the Quapaws' case. Hannum and Reverend J. F. Schermerhorn, secretary of the Stokes commission charged with addressing the problems of all Indians west of the Mississippi, met with the Quapaws on May 11, 1833. Hannun and Schermerhorn told the assembled group of the government's desire to move the tribe to western lands reserved for Indians. Although wanting to remain in Arkansas and fearing the extreme cold of winters in the promised western home, Chief Heckaton relented and signed a treaty on May 13 that called for the Quapaws' removal to 150 sections of land situated between the Seneca and Shawnee reservations west of the Missouri state line.[114] Following years of hardship and mistreatment by the United States Government, the language of Article 3 of the treaty must have seemed ludicrous to the beleaguered Quapaw. The treaty stated, "Whereas it is the policy of the United States in all their intercourse with the Indians to treat them liberally as well as justly, and to endeavour to promote their civilization and prosperity..." However, the treaty recognized the extensive cessions of land made by the Quapaw and their impoverished and wretched condition. The Quapaws received a considerable amount of farm implements, cows, hogs, sheep, oxen, blankets, rifles, shotguns, and miscellaneous other goods. Additionally, the government provided a farmer, a blacksmith, and a thousand dollars per year for educational purposes.[115] By any measure, a new life in Indian Territory should have been far superior to their tribulations in Arkansas and Louisiana.

Even though Chief Heckaton agreed in May 1833 to the tribe's removal to Indian Territory, a month later Chief Sarasin led approximately 300 Quapaws back to the Red River. It had only been six years earlier that the half-blood Chief Sarasin had led a quarter of the tribe back to Arkansas from among the Caddoes on the Red River. The hand full of Quapaws that remained with the Caddoes in 1827 had sent word to their Arkansas brethren that they would receive a part of the unpaid 1830 annuity if they would present themselves to the Caddo agent. The remainder of the 1830 annuity and an additional $1,000 were paid, but the agent rejected a request that the Quapaws be allowed to remain among the Caddoes.[116] Whether Sarasin actually made the trip back to the Red River in June of 1833 is in question. Sarasin, buried in the Catholic cemetery in old downtown Pine Bluff, rests under a tombstone inscribed: "Saracen, Chief of the Quapaws, died 1832" followed by "Friend of the Missionaries. Rescuer of captive children." Yet, in his December 1833 letter to the Secretary of War regarding the Quapaws' deteriorating condition as they waited for ratification of the treaty, Governor Pope added that, "Sarasen is old and superannuated and is led astray by every tale that is told him."

Given this written documentation that Sarasin was still alive, it is doubtful that Sarasin had died in 1832. One legend states that he accompanied a group of headmen to inspect the new lands in Indian Territory but went back to "sleep in the land of his fathers."[117]

The Treaty of 1833 signed by the Quapaws in May was finally ratified by the Senate almost a year later on April 12, 1834. On September 16, 1834, some three months after the majority of the tribe had left for Louisiana, Chief Heckaton and 175 remaining enrolled Quapaws gathered near Pine Bluff, Arkansas, and began their thirty-day trek to Indian Territory. The Quapaws settled in the northeast corner of Oklahoma, established villages, and in the spring of 1835 planted their first crops. In late 1835, fifty of the Quapaws who had gone with Sarasin to the Red River joined the Quapaws in Indian Territory. The remaining 250 refused to return and joined the Cherokees in Texas or took up temporary residence with the Choctaws just north of the Red River in Indian Territory.[118]

The 175 Quapaws who had moved to Indian Territory plus the remnant of Sarasin's band would face one more move. Based on a survey in 1836, the United States Government decided that the Quapaw should be located north of the Seneca and Shawnees.[119] This decision was made in spite of the fact that the treaty provided that the Quapaw would reside *between* the Seneca and Shawnees. The Quapaws did not learn of the survey until 1838 after considerable improvements had been made to the land now determined to belong to the Senecas. Although a move of only a relatively few miles compared to their earlier journeys, the Quapaws faced another upheaval in their tribal life in the fall of 1838, only four years after their arrival in Indian Territory. The Quapaws learned that they would not be paid for the improvements they had made on Seneca land even though the required move was due to a government mistake.[120] Additionally, the government would not let the Quapaws establish their traditional villages in the new reserved area but required the Quapaws to be scattered across the reservation on individual homesteads. Following these revelations, one hundred of the disgruntled tribe members left the reserve and joined their fellow Quapaws living with the Cherokees in northeastern Texas, who had recently moved from among the Caddoes on the Red River in Louisiana. Texas had become a temporary safe haven for the Quapaws under the leadership of Sam Houston, President of the Republic of Texas. Houston protected the tribes from eviction, even signing a treaty on February 23, 1836, with the Cherokees and associated bands (including the Quapaws) living in Texas. However, the Texas Senate refused to ratify the treaty in 1837. Houston's successor, Governor Lamar, sent troops to the Cherokee Nation to order their removal northward across the Red River into the Choctaw nation in July of

1839. Their refusal led to the massacre of 100 Cherokees including Chief Bowles. The evicted Quapaws did not return to the reservation in northeastern Oklahoma. After residing for a time in the Choctaw nation, the Quapaw refugees from Texas eventually moved north across the Canadian River.[121] In 1842, the Creeks allowed them to establish a village of approximately 250 residents up the Canadian approximately eight miles from Little River near present-day Holdenville.[122]

Back at the reservation, the Quapaws' efforts at becoming agriculturally self-sufficient were only marginal and somewhat undermined by the $2,000 annual annuity received between 1834 and 1854. When the funds began to be distributed on a per capita basis after 1840, many of the Canadian River Quapaws returned to the reservation once each year for the annuity distribution.[123]

After the final resettlement of the Quapaws in late 1838, the Indian agent in charge of the tribe prohibited the Quapaws from building and living in their traditional communal villages. As the Acting Superintendent of the Western Territories wrote in his annual report to Washington, "...the agent and farmer have wisely located each family to themselves, that they may cultivate separately" an effort to promote independent agricultural endeavors.[124] Although tribal life diminished with the government's efforts to change the traditional village system and the eventual allotments of their land, the Quapaws continued to maintain three basic tribal units with a chief presiding over each group, with one of the three being the hereditary leader of the tribe. After Chief Heckaton's death in 1842, War-te-she became the hereditary chief.[125]

The expiration in 1854 of the twenty-year annuity provided for by the Treaty of 1833 (ratified in 1834) ended the annual gatherings of the Canadian River Quapaws at the reservation. An even greater crisis arose as half of the 400 Quapaws abandoned the reservation and joined their Canadian River brothers living among the Creeks now that the annual annuity disbursement no longer held them to the reservation. With the end of the annuities in 1854, the Quapaws moved quickly to address the shortage of funds, and in November

1853, the tribe proposed a sale of its lands to remedy the looming crisis. Between 1853 and 1860, the Quapaws initiated and negotiated several proposals to sell all or parts of their reservation, all of which failed for lack of Senate ratification or denial by the Bureau of Indian Affairs. Likewise, the Quapaws rejected a government proposal that the reservation be surveyed and allotted to individual tribal members. With the organization of the Kansas Territory in 1854 and growing tension in both Kansas and the nation as the country inched toward the Civil War, the desire of the Quapaws to sell their land became a much lower priority than the larger and more serious problems facing the federal government.[126]

The Quapaws suffered from the ravages of the Civil War. In the spring of 1861, the Union withdrew its troops from the territory south of Kansas and abandoned the Indians, including the Quapaw to whom they were bound by treaty to protect. The Confederates capitalized on the withdrawal of the Union forces, and by August a faction of the Creeks along with the Choctaws, Chickasaws, Seminoles, Caddoes, Comanches, and Wichitas had signed treaties of alliance with the Confederate States. Soon to follow were the Cherokees. Cherokee Chief John Ross had been hesitant to abandon his neutrality and join the Cherokee Nation with the Confederacy. However, given the recent defeat of Union forces at Wilson's Creek in southwest Missouri, the reluctant chief decided to cast his lot with the Confederacy.[127] The Cherokees, along with the Quapaws, Osages, Senecas, and Shawnees, signed the Confederate treaty on October 4, 1861. In return the Quapaws were to receive $1,000 annually for educational purposes, a $2,000 annual annuity for twenty years and the construction of a gristmill.[128]

The Canadian River Quapaws joined Creek Chief Opothleyahola and 7,000 Creek and Seminole neutrals in refusing to recognize the treaties. Several battles were fought against Confederate Indian troops during the last two months of 1861. However, a defeat at the battle of Chustenalah on December 26 forced the neutrals including the Canadian River Quapaws to retreat northward to Leroy in present-day Coffey County located in east central Kansas. A census of the Opothleyahola refugees in southern Kansas that had survived defeat and endured the harsh winter march north reflected a population of 5,600 Creeks, 1,000 Seminoles, 140 Chickasaws, 315 Quapaws, 197 Delawares, and approximately 300 other Indians from various tribes.[129] Fearing advances of southern forces toward the reservation, the Quapaws in northeastern Oklahoma joined the Canadian River Quapaws near Leroy in February 1862. The tribe was reunited for the first time in twenty-seven years.[130] Other Quapaws were reported to be staying with the Shawnees near Lawrence, Kansas. Still others arrived at the Osage Mission where some of their children boarded while others settled with the Ottawas.[131]

Chief Wah-te-she died in January 1865 and was replaced by Ki-he-cah-te-da. Following one aborted attempt to return to their northeastern Oklahoma reservation, the Quapaws remained in Kansas until November 1865 even though the war had ended in April 1865. The Quapaws and other rebel tribes who had signed treaties with the Confederates were summoned to a meeting with Commissioner of Indian Affairs D. N. Cooley at Fort Smith on September 8. The Commissioner threatened the tribes with loss of their assigned lands because of their treaties with the Confederacy. But once the Commissioner understood the fact that federal troops had almost immediately withdrawn at the beginning of the war from the territory they were bound by treaty to protect, that most Quapaws had languished in refugee camps under great deprivations during the war, and that some Quapaws actively fought in the war against the Confederacy, the tribe was only required to sign a brief agreement calling for immediate peace. The agreement was signed on September 14, and in November 1865, 265 Quapaws returned to their reservation abandoned four-and-a-half years earlier to find houses looted or destroyed, fences gone, and cattle stolen.[132]

Apart from the ravages of disease in the late seventeenth and eighteenth centuries, nothing in their known history compares to the tribal destruction experienced during the first thirty-five years of the nineteenth century. The period of 1824 through 1834 was particularly devastating with the loss of their Arkansas homeland and the irreparable harm caused by the division of the tribe.

The United States Government's policies in dealing with the Indians during the eighteenth and nineteenth centuries can be called many things, including callous, self-serving, unfair, contradictory, indifferent, and only occasionally concerned and benevolent. However, *paradoxical* may be a better description of the government's Indian removal policy. For the American government, the Indian had moved from being a menace to a nuisance and then to a suffering and helpless ward of the state. The removal policy originally designed to provide lands for white settlers in the east eventually became a legitimate method to save the Indians. Many in government believed that the only way to save the Indian was to remove him from those elements of a white society that were corrupting, particularly to the Indian. Additionally, many thought that separation would allow time for the Indian to deal with if not adopt the white concepts of private ownership of land and property, individual effort, and individual self-reliance. But due to their weakened condition, the removal policy was also destined to destroy many tribes as independent entities. Many factors contributed to this weakened condition including the inter-tribal warfare that raged on a large scale before any significant penetration of European influence, the low resistance to diseases transported from foreign shores, alcohol, dependence on government annuities, and the general inability to discern the value and future consequences of their treaties and agreements.

From the moment Columbus stepped foot on the Western Hemisphere the Indian world would never be the same. As civilizations and technology advanced in other parts of the world, it was inevitable that the Indians of the Western Hemisphere would eventually encounter explorers from other lands. The continent could not continue in a vacuum, sealed from contact with the outside world. With all of its injustices, the European-Indian experience in North America was far superior to that of the Spanish-Indian experience in Mesoamerica and parts of the southwestern United States or the experiences of aboriginal peoples in other parts of the world. Just as many Europeans came to exploit the riches of the hemisphere, so too did many come to escape religious and ethnic persecutions and to establish a better life for themselves. And just as there were many injustices perpetrated on the indigenous population, so too did the Europeans eventually deliver the Indian from the savage inter-tribal wars, torture, brutality, slavery, and cannibalism. But without doubt, many of the European and later American government's shameful dealings with the American Indian expose a dark chapter in this continent's and our nation's histories.

The Quapaw—White Man's World

"For we are strangers before thee, and sojourners, as were all our fathers..."[133]

Before the beginning of the Civil War, the Quapaws had lived for almost thirty years on their reservation in the northeastern corner of Indian Territory bordered by Kansas and Missouri. During the war the Quapaws were absentee owners while living in Kansas refugee camps with other tribes that had escaped the hostilities south of the Kansas border. George C. Snow was appointed as the Indian Agent for the Neosho Agency on March 23, 1865. Escorted by fifty armed Indian soldiers, he arrived at Tar Springs, the head of Tar Creek, in early June. Upon his arrival, Snow was aghast at the destruction caused by both Union and Confederate forces and recommended to the Indian Department compensation for the Indians' war losses.[134] Many nationwide issues were being dealt with as Snow and his soldiers surveyed the destruction; President Lincoln had been assassinated less than two months earlier, and it would be another month before some of those implicated in the plot were hanged.[135] Snow's request for compensation for war damages to the tiny, distant Indian reservations in Indian Territory under his charge was not a high priority with federal officials as the nation reeled from Lincoln's assassination, the South lay devastated, and victorious northern states sought revenge rather than reconciliation.

Following the war and re-occupation of their tribal lands in November 1865, the Quapaws erected and lived in crude shelters. Bureau of Indian Affairs Special Agent George Mitchell provided monthly rations, but those ended in June 1866.[136] Crops planted in the spring of 1866 were flooded, and the tribe suffered for lack of food and clothes. Agent Snow in his annual report of September 1867 stated that he authorized use of $600 for purchase of ploughs, harnesses, and corn meal. The money had been designated for the hiring of a farmer to help the Quapaws in their agricultural endeavors.[137]

In desperate need for cash with which to feed and clothe tribe members, the Quapaws resorted to efforts to sell a portion of their reservation lands. Unlike the response to such proposals in the 1850s prior to the Civil War, the United States Government was receptive because of its desire to relocate the tribes that obstructed white settlement in Kansas. In December 1866, Agent Snow was directed to send a Quapaw delegation to Washington for negotiations along with representatives of other tribes within the agency's area of supervision and representatives of the tribes to be resettled. In January 1867, Quapaw chiefs John Hunker, Ka-she-ka, and interpreter Samuel Vallier (Valliere) were selected by the tribe to conduct the negotiations. Hunker died on the way to Washington.[138] Agent Snow's request for expenses for the Quapaw men making the trip to Washington gives an insight to the Quapaw hardships of the time:

> The Quapaws were at more expense than they would have been had they not meet with the misfortune to have lost one of their number.
> Also when they started out it was very cold and since they were nearly naked they had to buy clothing before they could get on the road.[139]

Ka-she-ka, Vallier, and the other tribes' representatives arrived in Washington, and negotiations opened on February 9.[140] General James G. Blunt had been engaged to represent the Quapaws, Senecas, Mixed Senecas, and Shawnees. The well-respected former commander of Union troops in Kansas in the latter stages of the Civil War was an experienced representative of Indian claims against the government.[141]

After two weeks of negotiations, the Treaty of 1867, later known as the Omnibus Treaty, was signed on February 23. The treaty provided that the Quapaws would sell 7,600 acres of their reservation located north of the Kansas/Indian Territory border for $1.25 per acre. A second tract consisting of 18,522 acres and representing the western one-fourth of the reservation would be sold for $1.15 per acre. The Miami and Peoria Indians would occupy this tract of land. Upon ratification of the treaty, $5,000 was to be immediately paid to relieve the tribe's hardships. The balance of the sale's proceeds was to be invested at 5% with interest distributed semi-annually on a per capita basis. An educational annuity of $1,000 would be used for Quapaw children to attend the Osage Manual Labor School until a school could be built on the reservation. In addition to other minor payments, a major provision of the treaty as far as the Indians were concerned was the establishment of a commission to adjudicate $90,000 in claims for losses suffered during the Civil War by the Quapaws, Senecas, Mixed Senecas, and Shawnees. It is believed that this remarkable concession was obtained by the tribes because of Blunt's influence with the government. However, Senate ratification of the Omnibus Treaty would not occur until June 18, 1868, due to a fifteen-month preoccupation with impeachment proceedings against President Andrew Johnson. The treaty including various amendments adopted by the Senate was sent to the tribes for ratification. However, the amended version eliminated the $90,000 request for compensation for wartime losses but provided for the appointment of a commission to investigate the claims and report directly to Congress. Strongly objecting to the loss of damages resulting from the war, the tribes nevertheless agreed, and the treaty was proclaimed on October 14, 1868.[142]

Following the treaty proclamation in October, an investigative commission was appointed and began its work on the war damage claims in March 1869. Documented losses to the four tribes totaled $110,809 of which $48,601 represented losses to the Quapaws. Through Blunt's continued efforts, Congress allocated $90,000 in July 1870 to settle the claims. During the investigation Agent Snow worked diligently to substantiate each claim for the commission. Additionally, as Blunt's representative, Snow secured from the individual Indian claimants an agreement to pay Blunt one-third of whatever the claimant was awarded as a fee. In October 1890, Blunt was paid $30,000, one-third of the claim. Two members of the Board of Indian Commissioners at the Neosho Agency to make payment to claimants vigorously protested Blunt's fee. Other high-level protests were made after the terms became publicly known and a congressional investigation labeled the fees as "extortion." But Blunt kept the money.[143] Iron-

QUAPAW CESSION—TREATY OF 1867.

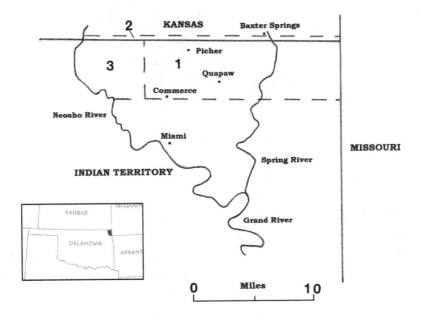

Tracts 1, 2, and 3 – Quapaw Reservation before Treaty of 1867

Tract 2: Strip of Land approximately one-half mile wide in Kansas and adjacent to the northern border of Indian Territory bound on the east by the Missouri border and the Neosho River on the west, containing 7,600 acres sold at $1.25 per acre.*

Tract 3: Beginning at the point where the Neosho River enters the northern border of Indian Territory, down the Neosho to point that the river intersects the southern border of the Quapaw reservation; thence east three miles along the southern border of the reservation; thence north to the northern border of Indian Territory; thence west along the northern border to the point of beginning at the Neosho River, containing 18,522 acres sold at $1.15 per acre.*

Tract 1 – Quapaw Reservation (After Cession)

*W. David Baird, *The Quapaw Indians*, (Norman, Oklahoma: University of Oklahoma Press, 1980), p. 107.

Figure 7

ically, fees for modern-day legal representation often exceed Blunt's requirement. Even if the one-third fee charged to the Qwapaws was comparable to modern-day fees, fees charged to other tribes by Blunt were not. For his counsel and influence in Washington, Blunt received $62,500 (50% fee) of $125,000 received by the Chickasaws. He received $33,600 (40%) of $84,000 received by the Choctaws. All totaled, Blunt received 42% of the $299,000 received by the three tribes. However excessive the fees may have been, without Blunt's efforts the Quapaw and other tribes almost certainly would have received nothing for their wartime losses.[144]

Government rations had been provided to the Quapaws during the winter of 1867–68, but the Quapaws fared no better during 1868 because of drought, crop failure, and loss of cattle from starvation.[145] The immediate payment of $5,000 due upon ratification of the amended treaty by the tribes in October of 1868 was not forthcoming, and the Quapaws were in desperate condition by March 1869. Many died that winter due to exposure, starvation, and alcohol. In the spring of 1869 Congress finally approved the appropriation, but the money was not received until fall of that year in spite of vigorous efforts by Agent Snow to speed the receipt of the funds by the suffering Quapaws.[146] Snow's disgust with the government's treatment of the Quapaws and other tribes under his charge was evident in his annual report of September 1869 to the Commissioner of Indian Affairs about the delay of the $5,000 payment to the Quapaws. Snow stated,

> "They (the Indians) have made many bitter, and I believe, just complaints against the government." In his report Snow berated the bureaucracy for their slowness to pay claims due under the treaty. He reminded his superiors that, "The superintendent received the funds in due time, put them in bank for safe-keeping, and they are there, safe, to this day, the superintendent being 'so pressed with other business' that he cannot make these payments in person, as he is required to do by law, and these poor wretches starving and begging for money due them…because the superintendent 'had not time to make the payment.'"[147]

Snow charged the government with blatant disregard for the provisions of its treaties with the Indians.

> When I view the failure on the part of the Government to comply with contracts made with a people who are considered capable of becoming 'parties to treaties', that have come under my own observation within the last seven years, it makes me wonder that we do not have more trouble with these benighted and ignorant people than we do.[148]

Snow ended his report by suggesting that the law creating Indian Agents should be repealed "so that 'Indian Agents' may no longer be the 'scape-goats' to bear the sins of the whole department."[149] Such sentiments must have gotten the attention of his superiors; this was his last annual report as United States Indian Agent for the Neosho Agency.

It would be 1872 before Congress appropriated funds for the ceded lands. Happily for the cash-strapped tribe, the $25,801 payment was made directly to the 240 Quapaws and amounted to over $100 per capita in lieu of the semi-annual interest payment of $2.69 per capita as provided by the treaty. The change in method of payment was in direct conflict with the provisions of the treaty.[150]

Even before their final departure from Arkansas and the 1833 treaty's provision of educational funds, the Quapaw saw the benefit of formal education for their youth. In 1830, at the request of Chief Heckaton, four Quapaw young men had been sent to the Choctaw Academy in Kentucky. The boys were still in school in 1836, but the superintendent threatened to send Rufus King and Washington Eaton to their homes because of their tendency to run away from school. However, the boys' behaviors improved and thus they were allowed to stay. By October 1838, both the Cherokees and Quapaws had refused to send any more boys to the Choctaw Academy.[151] However, two of the Quapaw boys remained in school and would not leave until 1841, over ten years after their arrival. One returned to the tribe but could not speak either Quapaw or English. The other was so confused that he could not find his way back to the Quapaws and made his home among the Omahas. A disgusted Chief Hecka-ton demanded that the $1,000 allocation for educational purposes be used to hire a teacher to teach on the reservation.[152]

In 1840, not waiting for action by the government, Heckaton asked Methodist minister Samuel Patterson from Sarcoxie, Missouri, to establish a school near the Quapaw reservation. Patterson's Quapaw Mission school began classes on March 27, 1843, on the east bank of the Spring River on the Quapaw reservation. The school was moved in 1844 to a new and more convenient location on the reservation five miles west of Newton County, Missouri, on the military road from Fort Leavenworth to Fort Smith. Renamed the Crawford Seminary in honor of Commissioner of Indian Affairs Thomas H. Crawford, the school was eventually expanded to accommodate forty students.[153] However, the school was closed in 1852 amidst a measles epidemic that killed forty tribe members.[154]

Missionary efforts by the Jesuits had been successful since 1847 and resurrected the latent Catholic sentiments present among the Quapaws since the French missionaries arrived at the Quapaw villages near the mouth of the Arkansas some 150 years earlier. The Roman Catholic Church had priests working among and befriending the Quapaws. Fifty-three Quapaws were baptized by the priests in 1850, a substantial portion of the small tribe. Even before the end of the Crawford Seminary, the Quapaws had applied for permission in 1851 to allow some of their children to attend the Osage Manual Labor School in Kansas sixty to seventy miles north of the reservation.[155] Chief War-te-she asked Jesuit Father Schoenmaker to minister to the education needs of the Quapaw children. So successful were the Jesuits' missionary efforts that 130 Quapaws were baptized between 1853 and 1855. As a result, many Quapaws gave up their traditional spirit-life beliefs.[156] The Quapaw school fund had been transferred from the Quapaw Mission School to the Osage Manual Labor School. The school term began with ten Quapaw children on February 28, 1853, and by the start of the term in September, the number had grown to eighteen boys and nine girls. By 1859, there were 130 Osage and twenty-two Quapaw children residing at the school.[157]

Following the Civil War, Father Schoenmaker's Osage Manual Labor School would continue receiving the Quapaws' annual education allowance, now provided by the terms of the Treaty of 1867. The federal government's payments to the Osage Manual Labor School ended in March 1870.[158] At about the same time, the Quakers were expanding their missionary work. As part of that work, they entered into a contract with the federal government to establish schools for the Indians located in the far northeastern corner of Indian Territory.[159] Hiram Jones, also a Quaker, was appointed as the resident agent for the Quapaw agency. Using government funds, Jones built a boarding school that opened on September 2, 1872, with Quaker teacher Emaline Tuttle, who had been transferred from the Ottawa school, and her husband who acted as superintendent. Average attendance never exceeded fifty at the school built to accommodate 100. The Quapaw mission school continued until 1879 when the couple retired. The school was reorganized and became the Quapaw Industrial Board School. Although attendance of the reorganized school averaged 112 at its peak, the Quapaw portion of the student body usually did not exceed twenty pupils. By the 1890s the campus included 160 acres with thirteen buildings and a farm. The government closed the school in 1900 as a means of economizing. However, the Quapaw Industrial Board School provided a strong academic education during its twenty-one years of existence; many of its pupils would become important leaders in tribal affairs during the first half of the twentieth century.[160]

The Indian schools were important not only for the education of Indian children but white children as well. Indian children had schools established in accordance with treaty provisions. Education for white children of the area was far inferior to that of the Indian children. However, from the beginning, the Indian day schools welcomed white children on a tuition basis. Whites had been renting Indian lands since 1881, and their numbers totaled 4,500 by 1894 with 5,000 acres under cultivation in the area of the Seneca, Peoria, Miami, Modoc, and Quapaw reservations and vicinity.[161]

Between 1873 and 1879, there was a persistent effort by Agent Hiram Jones and the Quaker couple at the Osage Mission School to get the Quapaws to allow settlement of other tribes on portions of the Quapaw reservation. Because of the Indian Agent's claims that the Quapaws had too much land and general disparagement of the tribe, some Quapaws began to move west into the Osage reserve in 1874—including Louis Angel, also known as Tallchief. Angel was the Quapaw hereditary chief who briefly succeeded Ki-he-cah-te-da upon his death in the fall of 1874. By 1876 about 115 Quapaws representing about half of the 235 tribe members had moved into the Osage reservation. The remaining reservation Quapaws totaled only seventy-five by the following year. As the number of reservation Quapaws declined, Agent Jones and the Osage Mission School couple continued pressure to have Quapaw lands assigned to other tribes; the remainder of the Quapaws moved among the Osages or to a more suitable location. Attempts were made to move several tribes into the reservation including the Modocs, Cheyennes, Comanches, Kiowas, and finally the Poncas in 1877.[162]

Agent Jones' annual reports to his superiors in Washington continually disparaged the Quapaws and sought to have their land taken for other tribes. In his first annual report in 1872, he reported there were 240 Quapaws on the reservation and claimed they were "the least developed of any of the tribes in this agency."[163] The 1873 report stated, "They have a reserve much larger than is needed for their use…"[164] In 1874, "I would suggest…that Government take a part of it for some other tribe…"[165] Again, in 1875, Jones' report stated, "The Quapaws seem to be the only exception to the general advancement, and they, I believe, have shown a better spirit this year than last; but a long course of indolence, vice, and idleness has so demoralized them, that I fear but little can be accomplished with the present adult generation."[166] The 1876 report cited a tribe population of 235 but stated, "At least one-half of these have left their reservation, and are living with the Osages…I would recommend that some action be taken as soon as practicable to dispose of their reservation to some tribe of loyal Indians…"[167]

Although the relocation efforts were promoted by Quaker Indian Agent Jones and the Quaker couple in charge of the Osage Mission School, it was another Quaker, Lewis Hadley, who came to the aid of the Quapaws. Hadley was hired at the school as a teacher in 1876 and assisted John Hotel and the other 102 reservation Quapaws in writing and presenting their grievances to the President and Commissioner of Indian Affairs. The correspondence indicated the Quapaws agreed to sell up to two-thirds of their reservation, but they adamantly refused to sell all of it or join the Quapaws residing in Osage territory. The proposal was never considered by Congress. For his efforts, Hadley was accused by Agent Jones of being unscrupulous and meddling and was fired from his teaching post. Because of his efforts on their behalf, the tribe appointed him as their council clerk and made him a member of the tribe with the name In-go-nom-pi-she, or One Who Can Not Be Put Down.[168]

The bearded and longhaired Hadley taught at several Indian schools and was a skilled linguist whose works included a Modoc dictionary and a long vocabulary of the Quapaw and Ponca languages. "He was thick-set rather than 'stout'—and had let his hair grow so as to be received among primitive Indians without suspicion…"[169] In any era Hadley would be considered as an interesting character. With his claim as an adopted citizen of the Quapaw tribe in dispute, Hadley wrote a letter in 1882 with regard to his experience.

> Mititi was probably the oldest and best looking woman in the tribe, had buried all of her family and wanting to marry. Your humble servt. was an adopted citizen fighting against certain schemes against the Quapaws, while the schemers were trying to dispute my rights as I was not maried (sic) into the tribe. Hence Mititi was working for my interest as well as her own, she knew I was in the market and sent me word she wanted to marry me. As I did not want a woman near a hundred years old I had to decline the tempting offer in such a way as not to give any offence.[170]

Perhaps more important than his actions on behalf of the Quapaws was Hadley's example in how to deal with government and other authorities impacting the Quapaws' lives. In-go-nom-pi-she's lesson would prove invaluable and would not soon be forgotten.

Jones' 1877 report stated that a majority of the tribe had long desired to remove to the Osage country and become a part of that tribe. He stated that such action was "bitterly opposed by a few (who were) backed by some unscrupulous intermeddling whites, who desire, for the advancement of their own interests, to thwart the wishes of the Government."[171] This was most likely a veiled allusion to Lewis Hadley's assistance to the reservation Quapaws. Jones' 1878 report would be his last, but again he claimed the Quapaws' were demoralized because of "their proximity to the vicious, intermeddling whites usually found on our border..."[172]

Troubles between the Sioux and Poncas led the government to send 170 Poncas to the Quapaw reservation based on misleading reports that the reservation was almost vacated and would be given to the Poncas. In June 1877, the first Poncas arrived and would remain until relocated in July 1878.[173] W. A. Howard, their appointed agent, found little to no preparation had been made for their arrival and informed the government that the situation was critical. Many of the Poncas were sick with malaria, and over one hundred would die while on the Quapaw reservation. The Poncas grew increasingly dissatisfied at staying on the Quapaw land without a treaty and proper title to the reservation. The attempt to permanently locate the Poncas on Quapaw land failed, but the government continued its efforts to have the Quapaws absorbed into the Osages and free the Quapaw reservation for use by other tribes.[174]

Encouraged by the attitude of the government officials, the Quapaws in Osage territory pressured their home band brethren to join them in 1882. Led by Chief Charley Quapaw, the home band resisted. Other efforts by the government included recommendation that the Tonkawas be placed on the Quapaw reservation, but again the Quapaws defeated those efforts. By 1889, not wanting to sell but due to the persistence of the government, the Quapaws offered to sell their land for $5.00 per acre and subsequently merge with the Osages. As an alternative, the Quapaws offered to accept an allotment of 320 acres to each member of the reservation, making the remaining land available for lease to either Indians or non-Indians. The Quapaws knew the demands were unrealistically high. As expected, the Bureau of Indian Affairs ignored the offer.[175] The government's attitude is not surprising given the fact that Congress had approved in late February 1889 the purchase of the ceded lands of the Creeks and Seminoles in the area known as Oklahoma country for less than a dollar per acre.[176] Perhaps following Lewis Hadley's example, the tribe was learning the art of hard-nosed negotiations, a necessary attribute in a white-dominated society.

When substantial deposits of lead and zinc were discovered in nearby southwestern Missouri in the 1870s and 1880s, the Quapaws began to realize the potential value of the reservation lands. But it would be many years before the Quapaw would know of and benefit from the vast mineral riches that lay beneath their feet. However, there were other more immediate sources of income. During the 1870s the Quapaw realized that the reservation could provide more than just a home for themselves with a few acres for cultivation. White farmers began crossing the Kansas border and establishing homesteads. Several of the Quapaws began leasing their land to the white homesteaders in exchange for one-third of the crop. By 1879, over 600 acres were under lease, and the white tenants' lease payments yielded 200 acres of crops for the Quapaws. Fearing the practice of leasing their land to be deleterious to developing within the Quapaws the value of individual labor, the government began restricting the practice in 1879.[177] Most homesteaders wrongly assumed the government had purchased the land when it was abandoned by the Ponca Indians after the failed attempt to settle the tribe on the Quapaw reservation. However, the lease system continued in a reduced and, on occasion, unauthorized fashion until the late 1880s.[178]

In addition to leasing their land to homesteaders, the Quapaws had another source of income that pre-dated their sharecropping enterprise. As early as 1868 the Quapaws were charging rent at the rate of one cow per month for each one hundred cattle grazing on the reservation.[179] With the advent of cattle drives from Texas to northern markets after the Civil War, the reservation became a convenient and desirable location to fatten the Texas cattle before sale in Kansas. The revenue from the cattlemen's use and crossing of the reservation grew from the late 1860s to 1879 when the government intervened to regulate the activity and protect the Quapaws' interests. The intervention was beneficial for the Quapaws and resulted in substantially increased fees and a more honest accounting from the cattlemen. Revenues continued to increase until 1885. Following the inauguration of Grover Cleveland as president in March 1885, the Bureau of Indian Affairs began a study of all non-Indian ranching on Indian reservation lands. As a result, the government cancelled all existing leases because it believed such activities tended to undermine the formation of the Indians' own agricultural pursuits.[180]

In the early 1880s fences had been constructed along the boundaries of the Quapaws', Ottawas', and Peorias' land that had been leased to cattlemen. Gates were installed where the boundary fences crossed the Military Road, the main supply road used by the Army during the settlement of the Indians, during the Civil War and now used by settlers in covered wagons headed west. The road entered the Quapaw reservation just south of Baxter Springs, Kansas, and entered the Cherokee Nation near Horse Creek. The road saw all manner of traffic, including the cattlemen leasing the reservation land, other cowboys, and an occasional outlaw running from territories where the presence of the law was more visible. The military presence in the territory since the Civil War to keep out white intruders was withdrawn in 1881. To provide for a measure of law and order, an Indian police force of eight men was organized.[181]

Special Indian Agent J. M. Hayworth reported in his annual report of August 27, 1879, that only thirty-eight Quapaws (including men, women, and children) remained on the reservation, all others having joined the Osage.[182] This number would drop to thirty-five in the agent's report of the following year.[183] By 1882, there were still

QUAPAW INDIAN AGENCY TRIBAL RESERVATIONS.

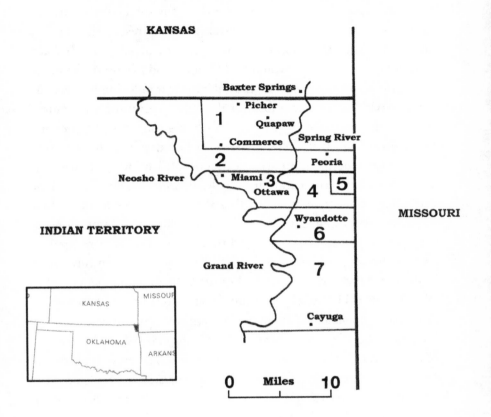

1. Quapaw
2. Peoria
3. Ottawa
4. Shawnee
5. Modoc
6. Wyandotte
7. Seneca

Figure 8

fewer than fifty home band Quapaws that remained on the reservation, which represented only about 20% of the total tribal membership. The number of Quapaws residing on the reservation had so diminished that the tribe began the practice of adopting Indians from other tribes if they would build homes and live on the reservation. This action was taken by the tribe for fear of losing its status as a nation with a subsequent loss of title to their reservation, a requirement of the Treaty of 1833. Paschal Fish, his son, and daughter-in-law, became the first non-Quapaws to be adopted into the tribe in 1880. Of Cherokee-Shawnee extraction, Fish built a house and allowed Kansas cattlemen to use the home as headquarters for their cattle operation, thereby permitting the cattle access to the entire, unfenced reservation grasslands. In April 1883, in spite of protests, the Secretary of the Interior ruled that the adoptions were legal so long as the adopted family resided on the reservation. In July, tribal leaders sent first councilman Alphonsus Vallier (Valliere) and council clerk Lewis Hadley to locate any Quapaw residing in their former Arkansas homeland and invite them to move to the reservation and join the tribe. The men traveled the Spring, Grand, and Arkansas Rivers by canoe and arrived in mid-August. The men visited several families. The trip bore fruit when three years later Dardenne and Abraham Ray arrived and were adopted by the Quapaw. Soon other Quapaw-French families began arriving.[184]

Following the arrival and adoption of their Arkansas Quapaw kinsmen, the tribe began a program of adopting members of other Indian tribes. The flood of non-Quapaw adoptions began in September 1886 and lasted until October 1887 and left a significant and lasting impact on the tribe. The wholesale adoptions stirred a number of protests from government officials as well as several absentee Quapaws still living among the Osages whose majority had not consented to the adoptions. Furthermore, there was strong evidence that three groups of adoptees had separately organized to gain eventual control of the reservation, its grasslands, and subsurface minerals. In March 1888, the Commissioner of Indian Affairs refused to recognize the adoption of a group of New York Indian extraction led by

A. G. McKenzie of Paola, Kansas. Eighteen members of the Miami tribe comprised a second group and were also denied membership in the Quapaw tribe.[185]

As a result of the adoptions, many of the absentee Quapaws living among the Osages began to see the monetary benefit of Quapaw citizenship, and a majority had returned to live on the reservation by 1887. The returning Quapaws gained political control of the reservation. In June 1888, wishing no further adoptions but seeking approval of the adoptions nullified by the March 1888 ruling, Chief Charley Quapaw, Frank and Alphonsus Vallier, and A. G. McKenzie traveled to Washington to appeal the denial by the Commissioner of Indian Affairs. The tribal representatives and McKenzie sought approval of both the McKenzie and Miami groups but opposed several other adoptions. The appeal did not result in approval of the adoptions but rather an investigation. In August 1888, Government Inspector Frank Armstrong visited the reservation. On February 2, 1889, following the assessment by Armstrong that most of the adoptions should be denied, the Secretary of the Interior ordered a Quapaw roll be established to include only the approved adoptions. In March 1889, two months before the approval of the final roll by the Secretary of the Interior, Abner W. Abrams acted for the council and hired a Washington attorney to push for recognition of the New York Indian adoptions but to oppose the inclusion of the eighteen Miami Indians on the Quapaw roll.[186]

The Secretary approved 116 names of enrolled members on May 27, and in June payment of grazing receipts totaling $4,500 was distributed on a per capita basis to tribal members. Efforts at reinstatement of some of the Indians to the Quapaw roll began almost immediately. Abrams and his family were included in the 116 members listed on the May roll. However, based on subsequent recommendations and claims that four full bloods had been mistakenly left off the roll caused the commissioner to order a second investigation. The investigation resulted in the issuance of a revised roll on October 18, 1889. Abrams and Alphonsus Vallier traveled to Washington in December and presented to the Commissioner of Indian Affairs a revised and corrected roll containing 205 names that had been approved by the tribal council. After several deletions, the Secretary approved a final roll of 193 names on February 8, 1890.[187]

The driving force behind the tribe's efforts with regard to adoptions was Abner Abrams. Adopted by the Quapaws on July 6, 1887, forty-year-old Abrams was a member of the New York Indian contingent led by A. G. McKenzie. Educated and intelligent, Abrams was soon appointed as clerk of the council of tribal leadership.[188]

One historian presents a different version of how Abrams came to be adopted by the Quapaws. Benjamin Tousey—a full-blood member of the Stockbridge tribe, one of the New York tribes—was living with the Senecas, and in talking to the Quapaws learned of their imminent loss of their reservation. Tousey suggested that they persuade his influential nephew, a veteran and hero of the Civil War, to move to the reservation and assist with retaining their lands. A group of headmen of the tribe made the first of several trips up the military road to Fulton, Kansas, near Fort Scott to talk with Abrams. Receiving a warm welcome from Abrams, the men were fed and entertained for several days. Abrams agreed to come; upon his arrival, the Quapaws presented the best house on the reservation for Abrams and his family's use. After the adoption of Abrams, his family, and several relatives, Abrams began a major reorganization of Quapaw business. Given the course of events and his domination of tribal government, it is believed that Abrams may have planned the original talks between Tousey and the Quapaws.[189]

By the summer of 1889, J. V. Summers had been the Indian Agent for the Quapaw Agency for four years. He had opposed the adoptions by the Quapaws but appeared to have had a change of heart. In his August 20, 1889, annual report, Summers stated,

> The Quapaws have shown decided signs of improvement during the year just ended. I attribute this largely to the recent adoptions by the Department of thrifty and industrious people amongst them…In this connection I would state that, notwithstanding my former objections to adoptions by the Quapaws,

which may be seen in my various reports thereon, I have become satisfied that adoptions...are beneficial to the tribe.[190]

It is not known whether Summers truly had a change of heart regarding adoptions or felt the political winds blowing from Washington following the Secretary of the Interior's approval of some of the adoptions.

Amidst all of the efforts to finalize the roll in accordance with the tribal council's wishes, the tribe was active in lobbying Congress for compensation for Indian Territory land granted by the Treaty of 1833 but never assigned. The Treaty of 1833 granted the Quapaws 150 sections (96,000 acres) that were reduced by the cession of 26,123 acres under terms of the Treaty of 1867. However, the Quapaws had only 56,685 acres in their reservation and wanted to be paid for the missing 13,192 acres.[191] In March 1890, the tribe hired former Kansas Governor Samuel Crawford, successful in having legislation introduced, to petition that would allow the Quapaws to bring legal action in the United States Court of Claims. However, Congress failed to act on the measure. Crawford's continued efforts and substantial influence resulted in approval of a direct appropriation for payment of the Quapaws' claims. The appropriation bill was approved on March 3, 1891, and allocated $39,575, or $3.00 per acre for the missing 13,192 acres. The legislation provided that $30,000 was to be distributed on a per capita basis with the remaining $9,575 to be paid to the tribe for disbursement in accordance with the council's direction. For all of his efforts Crawford received a fee of only $2,770, considerably less than General Blunt's fee of $30,000 twenty-one years earlier.[192]

Although opposing additional adoptions throughout 1891, on June 8, 1892, Abrams presented to the Commissioner of Indian Affairs a typewritten document signed by twenty Quapaws requesting adoption of seventeen persons into the tribe. As he had done over two years earlier when he approved the last roll of 193 names, the Secretary of the Interior approved the additional adoptions on June 29. When reports of bribery surfaced following the approval, an investigation by Special Agent Robert A. Gardner in October

1892 revealed that the tribal council had previously denied adoption of the seventeen applicants and that Abram's book of council proceedings did not reflect council action removing the prior objections. A vote of the Quapaws conducted by Gardner with regard to admission of the seventeen adoptees resulted in an overwhelming rejection of the adoptions. The special agent recommended rejection of the adoptions but failed to find evidence of bribery. Nevertheless, the Secretary of the Interior approved the adoptions bringing the Quapaw roll to 215 on February 24, 1893.[193]

Life had been hectic for the Quapaw leadership during the late 1880s and early 1890s because of the pressures from the government to sell their lands, the matters of adoption, and claims for land assigned but not received. However, the question of allotment added to their agenda and far exceeded the other matters in importance. The Dawes Allotment Act of 1887 provided that reservation lands were to be divided among the Indians residing on those reservations. Each head of a family would receive a 160-acre allotment. Each single person over eighteen and orphaned child under eighteen would receive an eighty-acre allotment.[194]

Dissatisfied with the 160-acre limit among other objections to the provisions of the act, the tribe sent Abner Abrams and Alphonsus Vallier to Washington in November 1889 to seek approval from Congress for special legislation that would allow each Quapaw tribe member to receive a 200-acre allotment. Samuel Crawford, who would later assist the tribe in its claims for assigned lands never received, was engaged to lobby Congress, but Crawford and the Quapaws had no success with the fifty-first Congress. In March 1892, almost two and one-half years after his first trip to Washington, Abrams presented forceful testimony to the Senate Committee on Indian Affairs during the second session of the fifty-second Congress. Abrams explained that the tribe was fearful that the 160-acre limit would leave much of the reservation un-allotted and subject to forced sale by the government at a price not agreeable to the Quapaws. In May and July respectively, the Senate and House favorably reported the bill out of committee. But the testimony and lobbying by Crawford proved fruitless as the fifty-second Congress ended without action on the bill.[195]

Given the expectation of approval of the 200-acre allotment for each tribe member, many Quapaws had selected their sites, constructed homes and other improvements, and began farming the land. Others had leased their land to white farmers. On March 23, 1893, in danger of losing their homes once again because of government mandate, the Quapaws took one of the most important actions in the tribe's modern history. Every adult tribe member with the exception of four gathered and unanimously agreed to allot the reservation without prior government approval. A committee was appointed to make the allotments. Heads of families would choose desired sites, and the council would settle any disputes that may arise. Apart from the individual allotments, 400 acres was set aside for a school and forty acres for the Catholic Church. A copy of the tribe's action was sent to the Bureau of Indian Affairs on April 6, and the allotments were substantially completed by August 1893. However, the reservation still contained 12,000 acres that had not been allotted; the Quapaws feared losing this portion of the reservation at some point in the future. Indian commissioners Henry L. Dawes and M. H. Kidd had come to Muskogee to negotiate the Five Civilized Tribes' allotments. In February 1894, both commissioners suggested that the remainder of the Quapaw reservation be allotted to the Quapaws. The following month the tribal council approved an additional allotment of forty acres to each tribe member listed on the roll. This second allotment was completed by year's end.[196]

With the help of Samuel Crawford and Oklahoma territorial delegate Dennis Flynn, the Quapaw efforts that once were expended on getting approval for 200-acre allotments for each tribe member were now spent on winning approval of the tribe's unilateral actions. Legislation was introduced by Delegate Flynn early in the second session of the fifty-third Congress that began in December 1893. However, it would be March 2, 1895, before Congress would approve an appropriations bill that ratified the actions of the Quapaws with regard to the tribe's allotment plan. The legislation provided that the allotments would be inalienable for twenty-five years from the date of patent and surplus lands on the reservations, if any, may be allotted to its members. The amendment inserted in the legislation mistakenly referred to the tribe and their council as the "Pawpaw" and the "Pawpaw National Council." Because the Secretary of the Interior had been given a supervisory role by the legislation, the Bureau of Indian Affairs was directed to investigate the procedures used by the tribe to make the allotments. Agent A. W. Able found evidence of bribery and favoritism in the allotment process and that Abner Abrams was responsible. Able reported that some of the persons included on the roll were undeserving while others were excluded without good reasons.[197] In one instance, Abrams had set aside the adoption of a Peoria family. The attempted eviction caused the death of Joseph Big Knife, also a Peoria Indian, who was shot and killed by Amos Vallier (Valliere) as he attempted to enforce the eviction. Sentenced to ten years in Leavenworth prison, Vallier was pardoned after two years.[198]

Political pressure instigated by Flynn resulted in the recall of Able and the end of the investigation. These actions only caused the conflict to intensify as the excluded Indians continued to press their claims through their attorney. It took the combined efforts of the Commissioner of Indian Affairs and Samuel Crawford as well as statements of the Quapaw council to overcome the weight of evidence reflecting the improprieties of the allotment process conducted by the Quapaw council. On March 28, 1896, the acting Secretary of the Interior approved the schedule of allotments as had been submitted by the Quapaws. Additionally, the Secretary ruled that no further adoptions would be allowed and that the fraud charges had not been proven. The 200-acre fee patents were issued on September 26, 1896, to 234 tribe members. The additional forty-acre fee patents were issued to the same tribe members on October 19 plus two additional patents to two members receiving only the 40-acre patent.[199]

In the sixty years that had elapsed since the arrival in Indian Territory, a remarkable transformation had occurred among the few

Quapaws who tenaciously hung on to their reservation. The tribe had repeatedly negotiated treaty after treaty in their quest for annuity. Sale of part or all of their lands was proposed numerous times by the Quapaw. The majority of the tribe had scattered itself to several other distant locations to the extent that by 1880 only thirty-five Quapaws remained on the reservation, less than 20% of the total Quapaw population. But by the 1870s, the few remaining reservation Quapaws began to realize the value of keeping the land. Private ownership of property that had been the antithesis of Indian life and thought for millennia was cast off as the Quapaw realized private ownership was necessary for survival. Additionally, they had become agriculturalists to a degree either through their own efforts or through leasing to white farmers. Christianity through the efforts of the Catholic Church had become well entrenched in tribal life. A belief in the value of formal education, initiated by the efforts of Reverend Samuel Patterson, would strengthen throughout the remainder of the nineteenth century. As one writer has stated, "The American Indian, it was said, was now on his way to becoming the 'Indian American.'"[200] The Quapaws had most certainly cast their lot with this group.

And even more remarkable were the drastic changes that occurred in the thirty years following the Civil War. Returning from the depredations of four years in refugee camps in Kansas during the war, the Quapaws were virtually without food and material goods of any consequence. All they had was the land. Through their fierce will and the help of a number of whites such as Father Schoenmaker, Indian Agent Snow, General James Blunt, Quaker Lewis Hadley, Governor Samuel Crawford, Territorial Delegate Dennis Flynn, numerous other government agents and friends of the tribe, the Quapaw survived and began to prosper. And in a few short years following the arrival of the twentieth century, some of their members would be counted as some of the wealthiest people on the planet.

Some modern-day historians and sociologists decry the loss of cultural identity and efforts to "civilize" the Indian. However, the Indian culture—like all cultures—does not exist in a vacuum. Well before the white man arrived on eastern shores, the cultures, beliefs, and pattern of life of the various tribes of the Ohio Valley had undergone numerous changes, both evolutionary and revolutionary, as had Indian tribes in other parts of the continent. The many changes were not painless or fair in many respects, either before or after the arrival of the white man. If the changes had been painless and fair, it would have been a unique moment in human history. Given the typical treatment of aboriginal societies throughout history, the Indians fared better than most, particularly after the founding of the United States.

Indian Country to Statehood—Life on the Edge

"Else if ye do in any wise go back, and cleave unto the remnant of these nations, even these that remain among you, and shall make marriages with them, and go in unto them, and they to you: Know for a certainty that the Lord your God will no more drive out any of these nations from before you... [201]

The very weather of Oklahoma was a reflection of the edginess of life in Indian country prior to statehood. The junction of three climate zones—the Dry Steppes climate of the southwestern U.S., the Humid Continental climate of the northeastern U.S., and the Humid Subtropical climate of the southeastern U.S.—caused the early settlers to build almost as many cellars as the ubiquitous outhouse. Situated in the center of these colliding air masses, the resulting spring and fall turbulence, separated by dry, scorching summers, often resulted in the spawning of tornadoes, torrential rains, and flooding. A bright, mild winter's day could end with subzero temperatures, blowing snow, and blizzard conditions. Just as quickly, a bright sun and gentle southern breeze would send thoughts of winter packing and tickle the senses with hints of springtime. Unpredictable as the weather and often as ferocious, Indian country was a boisterous frontier. During the last two decades of the nineteenth century and first decade of the twentieth century, this caldron began to boil—a caldron containing society's Indian outcasts and their tribal factions, cattlemen, farmers, politicians, opportunists, immigrants, businessmen, outlaws, confidence men, the poor, young families seeking to carve out a life, failures seeking a fresh start in life, and a sprinkling of Indian agents and lawmen sworn to bring some semblance of law and order to the region.

Between 1875 and 1896, Judge Isaac Parker held sway over the U.S. Court for the Western District of Arkansas located in Fort Smith. His jurisdiction included the 75,000 square miles and 60,000 inhabitants of Indian Territory. The "Hanging Judge" provided a semblance of the "law" half of law and order in Indian Territory that lay just to the west of Fort Smith and across the Arkansas border. Attempts at the "order" half of the equation was provided by Parker's 200 marshals who roamed the territory. A third of the lawmen would die in the line of duty as they rounded up the 9,000 lawbreakers sentenced during Parker's twenty-one year tenure on the bench at Fort Smith. One hundred and sixty lawbreakers would be sentenced to death by hanging, but Parker's hangman, a Bavarian immigrant named George Maledon, would actually hang only seventy-six. The territory was a haven for infamous desperados such as the James Gang, Belle Starr, Blue Duck, Ned Christie, Cherokee Bill, the Doolins, the Daltons, the Verdigris Kid, Henry Starr, and a bullwhip-wielding lady bandit called the Nighthawk Rider. The crime and lawlessness of the era peaked during the 1890s and then slowly declined. The last enclave of the outlaws was to be the northern portions of Indian Territory, areas occupied by the Cherokee, Osage, Quapaw, and other small reservations.[202]

The land that was to become known as Oklahoma did not hold the interests of the first European explorers. To the French and Spanish, the eastern forests and grassy plains did not promise the quick riches sought for their distant kings. Ultimately, all of Oklahoma except the panhandle would be acquired from France through the Louisiana Purchase in 1803.

The Northwest Ordinance of 1787 set out terms for government of the territory northwest of the Ohio River. The Ordinance provided a step-by-step process that became a pattern for admission of new states and granting of equal status. The Ordinance provided for a territorial government, representation in Washington, freedom of worship, right to trial by jury, public support of education, and interestingly, an early foundation with regards to the nation's attitude on the question of slavery.[203] Article 6 of the Ordinance provided, "There shall be neither slavery nor involuntary servitude in the said territory..."[204] The Northwest Ordinance was a radical departure from the accepted norm of the times of holding new territories as colonies with lesser status. Of all of the states carved from the Louisiana Territory, Oklahoma would be the last to become a state.[205]

Although the Northwest Ordinance was adopted only sixteen years before the Louisiana Purchase, it would be another 104 years after the 1803 acquisition before Oklahoma would become the forty-sixth state. In many respects, Oklahoma resembled a colony during that pre-statehood period as opposed to a state in the making.

The number of pre-historic Indian civilizations present in Oklahoma is unknown. However, it is known that prior to 1800, the Quapaw, Caddo, and Osage, along with the Plains Indians such as the Apache and Comanche, hunted over vast areas, including Oklahoma. But with Jefferson's idea of a permanent Indian area away from the rapidly expanding new nation coupled with the acquisition of the Louisiana Purchase, Oklahoma became an eventual dumping ground for over sixty Indian tribes during the 1800s.[206] Resettlement of distant tribes to Oklahoma began in the 1820s and lasted to the end of the 1880s. The adoption by Congress of the Indian Removal Act in 1830 resulted in the resettlement over the next fifteen years of 100,000 eastern Indians to lands west of Missouri and Arkansas into areas that became Kansas and Oklahoma. In 1854, other removals from the northern part of the territory had reduced Indian Territory to roughly an area bounded by the boarders of modern-day Oklahoma.

The Five Civilized Tribes owned most of this land until after the Civil War when the government forced the tribes to cede or sell parts of their lands because of their support of the South. The availability of these lands in western Oklahoma made possible the third and last removal between 1866 and 1885 when tribes from Texas, Kansas, and Nebraska were moved to Oklahoma. By 1885, all of Oklahoma had been assigned to various Indian tribes except a two million-acre area called the Unassigned Lands that lay in the center between Oklahoma and Indian Territories.[207]

The presence of white cattlemen and farmers in Indian country began to grow rapidly following the Civil War as various tribes leased their land to cattlemen for grazing and to farmers for growing crops. The land was also crisscrossed with wandering cowboys, outlaws, Indian agents, merchants, outlaws, and lawmen. Tension mounted between the cattlemen leasing vast acreages in Indian country surrounded by barbed wire and the landless farmers peering south over the Kansas border at virgin soil waiting for the plow. It would seem that the contest between the wealthy cattlemen and the poor farmer was a mismatch. However, the farmer had allies in the form of professional promoters called "Boomers" and those promoting railroad and town interests.[208]

Ironically, it was a mixed-blood Cherokee who had broken with his own tribe that brought national attention to the Unassigned Lands in the center of Indian country. Elias C. Boudinot, a lawyer that had represented certain railroad interests, argued persuasively but unsuccessfully before the Senate Committee on Territories for the establishment of an organized territory. One month after his January 17, 1879, appearance before the Senate, he wrote an article in the *Chicago Times* that in essence stated that two million acres known as "Oklahoma country" was unassigned to any Indian tribe and available for settlement.[209]

The Boudinot family was no stranger to strife with their fellow Cherokees. Preceding their forced removal to Oklahoma in 1838, two factions existed among the Cherokees. A one-eighth blood Cherokee of Scottish decent, John Ross was the Cherokee Nation's principal chief that led the faction opposing removal. Following the government's repeated failures to obtain an agreement from Ross for removal, the government turned to other influential Cherokees including leaders of the Treaty Party. The Treaty Party believed that the Cherokees' only hope lay in separating themselves from the whites and leaving their Georgia homelands. Elias Boudinot, editor of the *Cherokee Phoenix* and father of Elias C. Boudinot, Major Ridge (the full-blood speaker of the Cherokee National Council), and his son John Ridge were leading members of the Treaty Party that signed the Treaty of New Echota in 1835 deeding away their eastern lands and authorizing removal to western lands. But opposition to the treaty was widespread among the Cherokees, and by 1838 only two thousand of the seventeen thousand Cherokees had started westward. Major Ridge authored an earlier Cherokee statute providing for a death penalty to any Chero-

INDIAN TERRITORY LANDS—1866-1889.

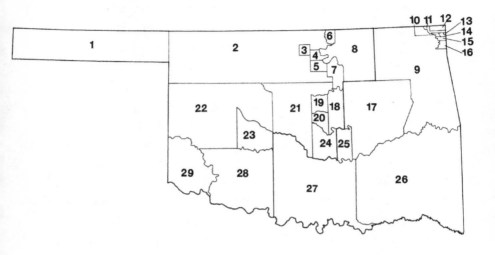

1. Land (unassigned)	17. Creek
2. Cherokee Outlet	18. Sac and Fox
3. Tonkawa	19. Iowa
4. Ponca	20. Kickapoo
5. Oto and Missouri	21. Unassigned Lands
6. Kaw	22. Cheyenne and Arapaho
7. Pawnee	23. Wichita and Caddo
8. Osage	24. Pottawatomie and Shawnee
9. Cherokee	25. Seminole
10. Ottawa	26. Choctaw
11. Peoria	27. Chickasaw
12. Quapaw	28. Comanche, Kiowa, and Apache No Man's
13. Modoc	29. Greer County
14. East Shawnee	
15. Wyandot	
16. Seneca	

Source: H. Wayne Morgan and Anne Hodges Morgan, Oklahoma, A History, (New York: W. W. Norton Company, 1984), p. 58.

Figure 9

kee deeding away tribal land. Yet, he signed the Treaty of New Echota due to his belief in the futility of resistance. Beginning in June 1838, a forced removal that became known as the infamous Trail of Tears was led by Major General Winfield Scott. More than four thousand Cherokees would die en route, including Chief John Ross' wife.[210]

The conflict between the Treaty Party and the Ross followers flared anew over leadership of the Cherokees upon their arrival in Oklahoma. Three of the four leaders of the Treaty Party, both Ridges and Elias Boudinot, were assassinated leading to a Cherokee civil war between 1839 and 1846. The tribal divisions would reoccur with the American Civil War and thereafter. Stan Watie, the only member of the Treaty Party to escape assassination, would become a Confederate general and the last to surrender at the close of the Civil War.[211]

Boudinot's 1879 *Chicago Times* article was the kindling needed to fuel the smoldering flame of white settlement in Indian country. Following the publication of Boudinot's article, the *Kansas City Times* began almost immediately to promote the invasion and settlement of Oklahoma country. The boomer movement was given legs by Captain David L. Payne, a former soldier and Kansas pioneer and politician. Payne was a commanding and gifted promoter. At six-foot-four and weighing over 200 pounds, he was described as very handsome with great oratorical skills. Vilified by some, proclaimed a great man by others, Payne was a colorful character with a common-law wife and free-spending ways. He had lost his farm to mortgage holders, worked in the House of Representatives rising to the post of Assistant Doorkeeper, and raised over $100,000 from his sales of memberships in the "Oklahoma Colony," a group of persons willing to personally invade and settle the Unassigned Lands. Payne led several failed attempts to settle homesteaders in Oklahoma country, lobbied politicians from local town meetings to the halls of Congress, and stirred significant press coverage until the Oklahoma country settlement had become a national issue.[212] Payne died at the age of fifty in 1884, but William L. Couch continued his efforts. However, it was the completion of the Santa Fe railroad from Arkansas City, Kansas, to Gainesville, Texas, in the spring of 1887 that gave hope to the boomer movement and those settlers frequently removed from Oklahoma country by the U.S. Army.[213]

This rail line through the center of Oklahoma country infused renewed vigor into the boomers' campaign to open Indian Territory to white settlement. Four men assumed the leadership of the boomer movement during a year of intense lobbying beginning in February 1888. In addition to W. L. Couch, Payne's successor, the group included Sidney Clarke, a former Kansas congressman; Samuel Crocker, editor of the *Oklahoma War Chief* newspaper; and Dr. Morrison Munford of the *Kansas City Times*. These men sent circulars to 500 men in Kansas, Colorado, New Mexico, Texas, Arkansas, Missouri, and Indian Territory inviting them to a meeting to be held on February 8, 1888, at the Kansas City Board of Trade Hall. In addition to numerous other speechmakers, Ottawa Chief John Earlie had been invited. The Ottawa tribe had already received their allotments, and Chief Earlie wanted the territory opened up. He believed that the Indians should mix with the whites, the whites should mix with the Indians, and the Indians should become more like the whites. The convention produced a resolution delivered to Congress by Munford, Couch, Clarke, and Crocker. Munford was a personal acquaintance with President Grover Cleveland and arranged a meeting of the delegation with the President.[214]

On January 6, 1888, William Springer of Illinois introduced his legislation in the House for the opening of the Oklahoma territory. On January 31, the House Committee on Territories chaired by Springer heard the protests of several Indian leaders and their attorneys. Debate in the House of Representatives commenced on February 25 and 28, 1888. The Springer bill would encounter numerous revisions, setbacks, failures, and resurrections over the next year in both the House and Senate.[215]

On February 27, 1889, with the approaching Senate adjournment and failure of the Springer bill, the annual Indian Appropriations Bill being considered in the house was amended to authorize the appropriation of $1,912,952 to pay for approximately 2.37 million acres of land ceded by the Creeks and Seminoles. A second amendment

OKLAHOMA TERRITORY LANDS—1889-1906 (Figure 10)

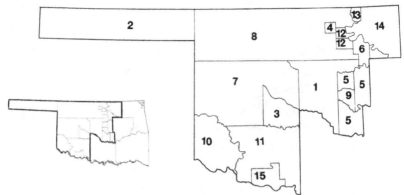

1. Unassigned Lands – Land Run – April 22, 1889
2. No Man's Land – Added to Oklahoma Territory by the Organic Act of May 2, 1890
3. Wichita and Caddo – Allotment agreement June 4, 1891; Lottery of surplus land   – June 9-August 6, 1901
4. Tonkawa – Allotment – 1891; Cherokee Outlet Land Run – September 16, 1893
5. Iowa, Pottawatomie, Sac & Fox, Shawnee – Land Run for surplus land– September 22, 1891
6. Pawnee – Allotment – 1892; Cherokee Outlet Land Run – September 16, 1893
7. Cheyenne & Arapaho – Allotment 1890; Land Run for surplus land – April 19, 1892
8. Cherokee Outlet – Land Run – September 16, 1893
9. Kickapoo – Land Run for surplus – May 23, 1895
10. Greer County – Supreme Court added to Oklahoma Territory on March 16, 1896
11. Apache, Comanche, & Kiowa – Allotment agreement October 6, 1892; lottery for surplus land – June 9-August 6, 1901
12. Ponca, Kansa, and Otoe-Missouri – Allotment 1904. No white settlement.
13. Kaw – Allotment – 1906. No white settlement.
14. Osage – Allotment – 1906. No white settlement.
15. Big Pasture (Apache, Comanche & Kiowa) – Sealed bid auction – December 1906

John W. Morris, Charles R. Goins, and Edwin C. McReynolds, *Historical Atlas of Oklahoma*, 2nd ed., rev. and enlarged (Norman, Oklahoma: University of Oklahoma Press, 1976), pp. 48, 50-51.

Odie B. Faulk, Oklahoma – Land of the Fair God, (Northridge, California: Windsor Publications, Inc.), pp. 129-132.

Anne Hodges Morgan and H. Wayne Morgan, ed., *Oklahoma – New views of the forty Sixth State*, (Norman, Oklahoma: University of Oklahoma Press, 1982), pp. 6, 38.

H. Wayne Morgan and Anne Hodges Morgan, *Oklahoma – A History*, (New York: W. W. Norton & Company, 1984), p. 58.

INDIAN TERRITORY – ALLOTMENTS &
PUBLIC AUCTIONS   -1889-1906 (Figure 11)

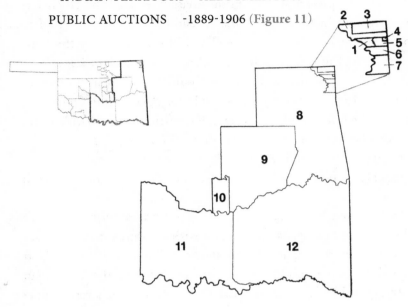

1. Ottawa – Allotment 1891

2. Peoria – Allotment 1893

3. Quapaw – Self allotment 1893, Approved 1895 – Allotment & Public Auction – June 1898-March 1907

4. Modoc – Allotment beginning 1890

5. East Shawnee

6. Wyandot – Allotment 1893

7. Seneca – Allotment 1902

8. Cherokee – Allotment & Public Auction – June 1898-March 1907

9. Creek – Allotment & Public Auction – June 1898-March 1907

10. Seminole – Allotment & Public Auction – June 1898-March 1907

11. Chickasaw – Allotment & Public Auction – June 1898-March 1907

12. Choctaw – Allotment & Public Auction – June 1898-March 1907

John W. Morris, Charles R. Goins, and Edwin C. McReynolds, Historical Atlas of Oklahoma, 2nd ed., rev. and enlarged (Norman, Oklahoma: University of Oklahoma Press, 1976), pp. 44-45. http://www.eighttribes. org/seneca-cayuga/ (accessed February 19, 2007).

Peoria Tribe Website: http://peoriatribe.com/history.php (accessed February 19, 2007).

Indian Land Tenure Foundations, Allotment Information for Easter Oklahoma BIA Region, http://www.indianlandtenure.org (accessed February 19, 2007.

authorized the president to open the unassigned lands by proclamation. Defeated in the Senate, the amendments were reinserted in a joint Senate-House conference. The bill was passed by the Senate and signed into law on March 2, 1889, by President Cleveland. Two days later, Benjamin Harrison would be sworn into office as the newly elected president, responsible for opening the Oklahoma country. The bill did not establish a new territory but provided that the ceded lands be opened to settlement.[216] On March 22, President Harrison gave the required thirty days notice of opening. The earliest possible entrance would be allowed and was set at noon on April 22, 1889.[217]

Washington in the spring of 1889 was crowded with Indians as well as Boomer lobbyists. Consumed with pressures from the government to sell their lands, the matters of adoption, claims for land assigned but not received, claims for losses during the Civil War, and the question of allotment, the Quapaws and their representatives were frequent visitors to Capitol Hill during 1889. It couldn't have escaped the Quapaws' attention that the success of the Boomer movement in opening the unassigned lands to white settlement would also portend dramatic changes for their future; this fact drove many of their lobbying efforts. Also, Ottawa Chief John Earlie's words of a year earlier at the Kansas City Boomer gathering had to have been remembered by the Quapaws. The chief's efforts to promote a merger of white and Indian societies must have been discussed and debated in many Quapaw cabins and council meetings.

Precedents for the Oklahoma land runs were events such as the opening of the Iowa farmlands in 1843 and 1845.[218] However, the Oklahoma land run of 1889 would not be like anything seen before. People sensed the American frontier was rapidly vanishing. For many the run represented their last chance to succeed in a country rushing headlong into industrialization, urbanization, and the looming twentieth century. Both the American and European press covered the event. Yes, there had been the California gold rush of 1849 followed by other dashes by those wishing to get-rich-quick on the precious metals found throughout the west. But the Oklahoma run was about land, homes, businesses, and the chance to settle and build a life. There would be the honest speculator and cheating Sooner staking claims for choice homesteads or town lots that could be turned for a quick profit. But these were a minority of the tens of thousands that lined the border of the unassigned lands straddling the heart of the future state.

On April 18, the Interior Department had allowed settlers to cross the Kansas border to traverse the sixty-mile wide Cherokee Outlet to the northern border of the Oklahoma territory. But the thousands of settlers heading for the northern boundary had to cross the flood-swollen Salt Creek fork of the Arkansas fourteen miles south of the Kansas boarder. Several settlers lost their possessions and some their lives trying to ford the stream before the railroad allowed one of its bridges to be used. The Boomers and soldiers worked all night covering the tracks with wooden planks. At 8 the next morning the first wagon and crossed, and by nightfall seven thousand men, women, and children had crossed with two thousand teams, mules, cattle, and horses. Another crossing would await them at Black Bear Creek.[219]

The 1889 dash for unassigned lands would be called Harrison's Hoss Race and would occur less than two months after outgoing President Cleveland signed the bill into law. Government surveyors had divided the two million plus acres into 160-acre homesteads and several larger town sites including two federal land offices for filing of claims, Guthrie Station where the Santa Fe Railroad crossed the Cimarron, and the Kingfisher stage stop.[220] The Oklahoma district covered a generally rectangular tract measuring approximately forty miles wide by sixty miles long with additional areas protruding from three corners. It was estimated that as many as 100,000 made the run on that cloudless spring Monday in April following Easter celebrations of the previous day.[221] Whatever the actual number, a mass of humanity lined the borders of the "Promised Land." The largest contingent was the northerners and easterners that entered through Kansas and lined the southern edge of the Cherokee Outlet. At each of the eight major crossing points, the line of hopeful landowners stretched to the horizon. Among them were farmers, laborers, mechanics, professional men, merchants, gamblers, confidence men, and an assortment of adventurers. Most were young men, but there were a few women, blacks, and immigrants. They rode fast horses, plodding plow horses, buckboards, surreys, buggies, wagons, oxen, donkeys, and an occasional bicycle. Some walked and others rode slow-moving trains. Fifteen trains entered the Oklahoma district from the north packed with passengers both within and on the roofs of the cars. Other trains would enter from the south at Purcell. Army buglers had been dispersed along the border. At one minute before twelve on April 22, the faint notes of distant horns could be heard followed by the sharp report of rifles fired by Army officers, and the stampede was on.[222]

The successful settlers claimed substantially all of the two million acres on that first day. Using two-foot wooden stakes, the settlers literally staked their claims to town sites or 160-acre homesteads. On April 23, the sun rose over Guthrie, Kingfisher, Norman, Edmond, Oklahoma City, El Reno, Stillwater, and several other new towns that sprang from the prairies and river banks where little existed at sunrise the day before except for quiet streams and grasslands.[223]

By evening of the run, Guthrie would see 10,000 to 15,000 camped in and around the new city.[224] Hundreds had been present at Guthrie prior to the run as officials, railroad workers, and others who managed to stay and grab claims and town sites. Thousands of other "sooners" had staked their claims well ahead of the official start of the run. One humorous story was told about two men on fast horses who arrived at a choice spot to find an old man plowing with a team of oxen. He claimed that he had gotten there only fifteen minutes earlier and that the soil was so rich, the four-inch high onions had grown that much since he planted them upon his arrival.[225]

Disputes among claimants would occupy the courts for years. However, many claims were ended with a ruling on October 1, 1890, by Secretary of the Interior John W. Noble who defined anyone entering the Oklahoma lands prior to the start of the run, with or without permit, as a Sooner and their claims would be defeated.

Other Sooners had based their claims on making the run from the railroad right-of-way while in Oklahoma lands. These claims were also overturned.[226] Other disputes ended in violence. Ironically, William Couch, Payne's successor as one of the leaders of the Boomer movement, was shot in the knee in April 1890 following a yearlong dispute over a prime piece of land with a Sooner. The shooter was J. C. Adams, the mayor of west Oklahoma City at the same time Couch had been the mayor of Oklahoma City. Couch developed gangrene and died. His assailant was sent to prison.[227]

There were many heartbreaking stories on the day of the run. Many lost their last hope of starting a new life. Some lost their lives or lives of family members. Many arrived too late or were cheated from obtaining a desired site by a Sooner and were forced to leave in defeat. However, those who were successful found considerable hardships that taxed their determination to stay on their land. Because of the difficulty in finding wood, many a homesteader resorted to the only type of shelter available: a sodhouse or dugout. Dirt walls could be covered with newspapers or plaster, but rats burrowed into the walls, allowing water to stream inside when it rained. Mosquitoes, bedbugs, and fleas made their homes with the residents of the usually overcrowded sodhouse or dugout. The drought in 1890 caused many homesteaders to return from where they came. Such a life filled with danger, depravation, monotony, sickness, worry, and loneliness was particularly hard on women. Many retreated to alcohol or opium. Depression and madness often preceded suicide. The danger encountered by these pioneers is illustrated by the story told of one woman who lay down on her bed in the afternoon to rest with her small child. She dozed but was roused somewhat when her child repeatedly peered over the side of the bed before pulling back with laughter. When she looked to see what was entertaining the child, she was horrified to find a coiled rattlesnake on the floor beside the bed. Each time the snake would attempt a strike, the child would pull back.[228]

In May 1890, Indian country was officially divided between Oklahoma Territory and Indian Territory. Indian Territory included the reservation lands of the Five Civilized Tribes and the small reservations of other tribes in the far northeastern portion of the territory. The remainder was Oklahoma Territory and include all reservations in the Cherokee Outlet except that portion still owned by the Cherokees. The government later purchased that portion in 1893.[229]

For those that missed staking a claim in the run of 1889, there were other opportunities as a series of land openings would occur prior to statehood in 1907. Additional land runs occurred in 1891 (Iowa, Sac and Fox, Shawnee, and Pottawatomie), 1892 (Cheyenne and Arapaho), 1893 (Cherokee Outlet), and 1895 (Kickapoo).[230]

Cattlemen had offered to purchase the Cherokee Outlet from the Cherokees for three dollars per acre if the tribe could secure approval for the sale from the government. As they were receiving rents of $200,000 annually from the cattlemen, they declined the offer. However, the government soon ordered all of the cattlemen out of the Outlet, thus ending the Cherokees' annual fee income. The tribe finally succumbed to the government's pressure to sell the Outlet lands and accepted $1.40 per acre in December 1891.[231] This prime piece of grassland opened to settlement in September 1893 in what would be known as the greatest land run of them all. By the day of the run on September 16, an estimated 100,000 to 150,000 people positioned themselves on the northern and southern boarders of the Outlet waiting to charge across the treeless prairie. Thousands of onlookers had gathered to watch the spectacle.[232]

Kansas and Oklahoma experienced a severe drought during the summer of 1893. No rain fell in August, and daily temperatures exceeded 100 degrees. Water sold for a nickel per glass to a dollar a bucket along the line of settlers waiting for the run as both water and food were in short supply. Living conditions for those wait-ing on the northern and southern borders of the Outlet resembled crowded, war-weary refugee camps. The cloudless sky was filled with dust kicked up by the teeming masses of people and animals. Soldiers set grass fires near the borders to allow them to better patrol the perimeter and guard against early entrants. The smoke from the fires added to the choking conditions. Many were overcome by the heat and died. Others perished because of a lack of water.[233]

As noon approached on September 16, a twelve-year-old Indian boy's report on the start of the run gives an understanding of the immensity of the crowds.

> There was a ridge about five miles in my lead that I wanted to reach before the race took place. On gaining the top of the ridge I was about ten minutes ahead of time. I found the people there just as wild as they were at other places. They were all standing now, and each one trying to get as close to the line as he could. I was situated where I could see fifteen miles to the east, and as far as my eye could reach to the west, and as far as I could see there were nothing but people, people by the thousands.[234]

Soldiers were posted every 600 yards to give the starting signal and to stop any who crossed the line before noon. A deaf man mistakenly thought the starting signal had been given moved forward and was shot from his horse by a soldier. Another galloped across the line thinking that shot signaled the start of the run. When he failed to heed commands to stop, he too was shot and killed by a drunken soldier.[235] Another tragedy occurred as an old man was having difficulty controlling a young, spirited horse. After jumping over the line once and being warned by a soldier, the horse jumped across the line again. The soldier raised his rifle and shot the old man just below his arm, killing him. The old man's son was fifty yards down the line. He saw the soldier kill his father and immediately drew his revolver and killed the soldier. The crowd took little notice of the two bodies placed in a wagon and covered with hay that had been brought for the horses.[236]

As the hot sun approached its zenith over the parched prairie, the official starting gunshots were fired. The two lines of seething, restless, hopeful pioneers thirty-five miles apart surged forward toward each other. The young Indian boy reported that he saw "...cripples and old women afoot, families in ox carts drawn by oxen, men on bicycles, in wagons, buggies, and men and women on horse-back..."

A buggy carrying two women turned over when their horse stepped into a badger hole. The women were thrown from the buggy, and the stampeded crowd rushed over them. The countryside was littered with broken conveyances and maimed livestock and humans alike.[237] All 40,000 homesteads available were claimed within two hours, and the largest and perhaps greatest race in world history was over.[238]

Because of the lawlessness and disorder of the land runs, two lotteries were simultaneously held in 1901, one for the Wichita and Caddo lands and the other for the Comanche, Kiowa, and Apache lands. Registration began on June 9 and by July 28, approximately 165,000 people had registered for one of the 1,500 homesteads available. The drawings occurred between July 29 and August 5. However, a part of the Kiowa and Comanche lands were not included in the lottery. Known as the Big Pasture Reserve, these lands would be sold by auction in December 1906 fetching an average of $10 per acre with no individual allowed to purchase more that 160 acres.[239]

Several reservations were being divided among the tribal occupants by allotment as had been decreed by the Dawes Allotment Act of 1887. The Ponca, Kansa, and Otoe-Missouri allotments occurred in 1904, the Kaw reservation and the large Osage reservation allotments in 1906. Excess lands not used for allotment in the Ponca, Kansa, and Otoe-Missouri reservations were opened for settlement between 1901 and 1906.[240] However, since the Osage had acquired the vast area to become Osage County from the Cherokees, no white settlement would be allowed. Consequently, each tribe member would be allotted almost a section of land, about four times the normal 160-acre allotment. More important, the tribe as a whole retained title to the mineral rights of their land.[241]

Oklahoma's boundaries were rapidly expanding. The strip of land west of Oklahoma Territory known as No Man's Land had not been made a part Texas, New Mexico, or Kansas. Robber's Roost was another name for the lawless strip of land that became a haven for outlaws following the Civil War. A few farmers and cattlemen petitioned Congress to organize the 160 by 35-mile area as the Territory of Cimarron. Their efforts failed, and the Organic Act of 1890 would add the Panhandle of Oklahoma to Oklahoma Territory.[242] Another addition occurred when Greer County was added. A dispute had risen with Texas over which of the two branches of the Red River separated Oklahoma from Texas. Oklahomans claimed the south fork was the main branch leaving Greer County as part of Oklahoma Territory. The Supreme Court sided with the Oklahomans on March 16, 1896, and Greer County officially became part of Oklahoma Territory. Texas homesteaders were allowed to remain on their 160-acre homesteads and file for an additional 160 acres at a cost of one dollar per acre.[243]

Although selling a majority of the Cherokee Outlet in 1891, a small portion had been retained until its sale in 1893. Thereafter, Indian Territory included only the lands of the Five Civilized Tribes and the small reservations in the far northeastern corner of the territory. In 1893, Congress created the Dawes Commission charged with obtaining agreements with the Five Civilized Tribes that they would allot their lands to individual tribe members.[244] These allotments would begin in June 1898. By March 1907, the rolls of the Five Civilized Tribes totaled 101,526 with more than half having less than one-half Indian Blood. A separate enrollment of 23,405 blacks, or freedmen, was made. Of the Five Civilized Tribes' 19,525,966 acres of land, 15,794,400 acres were allotted. The balance of the 3.7 million acres was sold at public auction and included 309 town sites and other un-allotted lands. Many Indians avoided enrollment while other full bloods enrolled as quarter bloods or less to avoid the appointment of guardians and thereby have less restrictions on their lands.[245]

With the official division of Oklahoma Territory and Indian Territory in 1890, the immediate push for statehood began. The first Oklahoma Statehood Convention was held in Oklahoma City in 1891. The convention prompted a bill in Congress that would have combined both territories as one state. The bill quickly failed but set the stage for several proposals during the 1890s. The debate on these proposals centered on single statehood or double statehood. By 1900, combining both territories into a single state had gained ascendancy among the majority of Oklahomans, but Congressional efforts to this end failed in 1902, 1903, and 1904.[246]

Indian Territory leaders were opposed to joint statehood with Oklahoma Territory and the resulting loss of tribal lands and governments. These leaders had become convinced that action was necessary to prevent the unification of the two territories, and on August 21, 1905, 182 elected delegates, both white and red, met at Muskogee. The Sequoyah Convention elected Creek Chief Pleasant Porter as its president. The five vice presidents representing the tribes included W. C. Rogers, father of famed humorist Will Rogers, and two future governors of Oklahoma, Charles N. Haskell and William "Alfalfa Bill" Murray, both non-Indians. The convention adopted a constitution that called for the creation of a state called Sequoyah. Haskell also secured an agreement from the delegates that if separate statehood for Indian Territory failed, there would be no opposition to joint statehood with Oklahoma Territory. The residents of Indian Territory voted to approve the proposed constitution on November 7, 1905, by an overwhelming majority of six to one.[247]

The following month a bill was introduced in the House of Representatives to admit the state of Sequoyah. Five other bills proposing single statehood for both territories were introduced during the debate. The bill to admit Sequoyah as a state failed, but on June 16, 1906, the Oklahoma Enabling Act was passed. The act provided for a constitutional convention with fifty-five delegates each from the two territories and two from the Osage nation. The convention was convened in Guthrie on November 20, 1906. The Democrats captured 100 of the 112 seats and elected "Alfalfa Bill" Murray as president. The convention completed its work and adjourned on March 15, 1907. Generally, the constitution produced by the convention gave the voter strength through initiative petition and referendum powers, limited the power of government, and maximized tax revenue from businesses while tightly controlling their activities.[248] It was Murray's belief that the new Oklahoma Constitution owed many of its most important provisions to the work done on the Sequoyah Constitution particularly those areas that dealt with corporate own-

ership of land and the protection of individual ownership and operation of farms.[249]

Murray did not immediately file a final copy with the territorial secretary, preferring to gauge public and presidential reactions.
President Roosevelt, using mostly colorful and unprintable language,
expressed his displeasure with much of the proposed constitution
and offered several suggestions. Murray reconvened the convention
and made some of the recommended changes. William Jennings
Bryan, an ardent foe of Theodore Roosevelt, would soon campaign
a third time for the presidency. Bryan strongly campaigned for the
adoption of the constitution while Roosevelt sent Secretary of War
William Howard Taft to Oklahoma to campaign against it, hoping
to force the two territories to produce a more acceptable document.
Nevertheless, on September 17, residents approved the constitution
with a 71% majority.[250] Democrats also swept to victory in the elec-
tive offices voted upon during the same election with Charles N.
Haskell as governor, a majority of both houses of the legislature,
most state offices, and four of the five Congressional seats. Not
only would Bryan win the battle for approval of the constitution, he
would win Oklahoma's first presidential election one year later in his
failed attempt to defeat Taft for the presidency.[251]

In spite of Roosevelt's opposition to the Oklahoma constitution,
the terms of the Enabling Act were met and Oklahoma became
the forty-sixth state on November 16, 1907, amid shouts of joy and
celebration. Brass bands, fireworks, picnic dinners, and celebrations
occurred in every town in the new state as businesses closed for the
day. But most of the older Indians could not be found among the
festivities. Official ceremonies were held on the front steps of the
Carnegie Library in Guthrie including the oath of office taken by
Oklahoma's first governor, Charles Haskell, and the symbolic wedding of an Indian Territory Cherokee bride and a young man from
Oklahoma Territory.[252] But the fact that the voters had also chosen
statewide prohibition meant that saloons would close for the final
time at midnight. This was a sobering thought for many and would

usher in an era that would last in some fashion for well over a half century.[253]

To understand the Indians' reactions to that first day of statehood, a story is often recounted in books on Oklahoma history of a Cherokee woman's refusal of her white husband's invitation to attend one of the many celebrations. Having gone alone, he returned to say, "Well, Mary, we no longer live in the Cherokee Nation." Thirty years had passed but the tears fell from her eyes when she recalled that day. "It broke my heart. I went to bed and cried all night long. It seemed more than I could bear that the Cherokee Nation, my country and my people's country, was no more."[254] How many other Cherokee Marys laid their heads on their pillows and cried through the night of November 16, 1907? This author's great grandmother, Mary Elzina Downey, was a Cherokee woman married to a white man who lived in eastern Oklahoma at that time; she must have been near the same age as the Cherokee woman quoted above. I shall always wonder what her thoughts and feelings were as she lay on her bed following that momentous day.

Although many Oklahoma Indians grew to be proud of their state and its heritage, the deep bitterness that developed over a century of mistreatment lingered for decades. While working on a school assignment in the 1950s, the author will never forget a great aunt's indignation and choice words at the mention of Andrew Jackson and his role in the Trail of Tears. Although nearly a century and a quarter had elapsed at that time since her Cherokee ancestors arrived in Oklahoma, the emotions and anger passed down from previous generations was still very much alive.

PART II—LEAD AND ZINC MINING ERA IN THE TRI-STATE REGION

What was it like to be in the Tri-State lead and zinc mining district in the latter part of the nineteenth century and first half of the twentieth century? There is not a single answer but many. "Wildcatters and Prospectors" recounts the early discoveries of lead and zinc in the Tri-State region of Missouri, Kansas, and Oklahoma and includes the almost simultaneous discovery of both oil and lead on Indian reservations in the northeastern portion of Indian Territory. "Boomtown!" is the story of the world's largest lead and zinc discovery on the Quapaw Reservation in 1914 and the founding of Picher, Oklahoma. In "Brother Lead and Cousin Zinc" the story digresses to the history of the ores and their uses and importance. "Into the Pits" tells how lead and zinc is mined and what it is like to be a lead and zinc miner. "Mountains of Chaff" continues the story of lead and zinc after it arrives at the surface. Life in the Tri-State mining district in the early twentieth century was not for the weak or faint of heart as we see in "Hell's Fringe," that portion of Indian Territory that lay just inside its border with Kansas, Missouri, and Arkansas. The area was populated by numerous characters and personalities who were famous and infamous such as Mickey Mantle and Bonnie and Clyde. Others were little know such as Uncle Charley, reported to be the last man killed in an underground accident in the Picher mining field. "Miners' Health and Labor Strife" describes the dangers of working and living in the mining district and the initial efforts at labor organization. "Pickets, Pick Handles, and Police" continues the story of labor unrest in the Tri-State Mining District that became known as one of the bloodiest chapters in the American labor movement.

Wildcatters and Prospectors

"...and the earth shall yield her increase..."[255]

The discovery of huge deposits of lead and zinc in the far northeast corner of Indian Territory in the early years of the twentieth century was big news but not nearly as big as the news of the discovery of oil a few miles to the west. The oil industry was born in the United States on August 27, 1859, at Titusville, Pennsylvania, when former railroad conductor Colonel Edwin Drake drilled the first successful well. The fledgling industry soon spread to surrounding states.[256] In August 1875, sixteen years after the Titusville discovery, an Indian cowboy and a white trader were rounding up stray cattle on an Indian Territory reservation occupied by 1,500 Osage Indians after their re-settlement from eastern Kansas three years earlier. The two thirsty cowboys guided their horses to Sand Creek, but the horses refused to drink because of the oily scum floating on the surface. Jasper Exendine, the Delaware Indian-half of the duo, pushed back the layer of scum that smelled like coal oil so the horses could drink. He berated its presence. "Oil no good. Make water bad." George B. Keeler, his white trader sidekick, said little but was in a thoughtful mood as the two rode back to the small hamlet surrounding Jake Bartles' trading post and gristmill on the banks of the Caney River. Keeler remembered the stories about oil and knew it had value other than for rheumatic muscles and joints, cuts on horses, and numerous other home remedies used by the Indians and frontiersmen.[257]

Keeler clerked for Bartles, but in 1884 he and a young transplanted Canadian named William Johnstone, another Bartles' employee, opened a store on the south side of the Caney River. Johnstone's wife was a niece of Bartles and daughter of Chief Journeycake. By 1897 the village had grown, prospered, and was incorporated as Bartlesville. Yet with all his prosperity as a storeowner, Keeler had not forgotten his experience at the oil-covered creek twenty years earlier. Keeler convinced Johnstone, Frank Overlees (who was another local storeowner), and several others to join him in searching for a commercial vein of oil. After securing a land lease from the Cherokee Nation to allow drilling, the men selected Cudahy Oil Company to develop the lease and drill several sites in Bartlesville. Drilling began only fifteen days after incorporation of the town. After several dry holes, the hopeful oilmen began drilling on the Cherokee allotment of William Johnstone's stepdaughter. After the well reached a depth of 1,303 feet, George Keeler's stepdaughter, Jenny Cass, dropped an explosive charge down the drill hole. The "go-devil" as it was called ignited the nitroglycerin to open the well and release any crude oil trapped below. Hardly had the noise of the blast reverberated across the prairie when a column of pure crude exploded up the borehole and shot high above the top of Nellie Johnstone No. 1's wooden derrick showering the owners and onlookers. That day, April 15, 1897, marked the beginning of the Oklahoma oil boom. Soon other derricks sprouted in Bartlesville and on the nearby Osage reservation whose eastern boundary was very near the newly incorporated town's western edge.[258]

Six years after the Nellie Johnstone No. 1, a successful young businessman—who would play a pivotal role in the Oklahoma oil boom—stopped in St. Louis to see the palaces and exhibition halls nearing completion for the soon-to-open World's Fair. Known as the Louisiana Purchase Exposition, it would mark the one hundredth anniversary of its acquisition. The successful young businessman was Frank Phillips, who was returning to Creston, Iowa, from a business trip to Chicago. Phillips' chance meeting with C. B. Larrabee, a Methodist minister from Creston, would change his life and the lives of thousands. Larrabee regaled Phillips with descriptions of the wild Indian Territory and his efforts to minister to its inhabitants. But it was the minister's description of Bartlesville and the fledgling oil industry that riveted Frank's attention and sparked his entrepreneurial spirit. "The oil is flowing out of the ground like water," said Larrabee. "It's liquid gold and men are getting rich. Very rich." Phillips forgot about the fair and could hardly wait to get back to Creston to tell his banker father-in-law what he had heard.[259] After a quick trip to see Bartlesville for himself, Frank enlisted his brother, L.E., to join him in his venture and move to Oklahoma. Meeting in Frank's Des Moines hotel room, Frank pointed to a passing automobile and said, "I think people are going to buy quite a few of these new buggies and need gasoline to make 'em go. It may be the thing of the future. There might be only a few gas wagons now, but someday there will be millions of the things."[260]

Before that eventful year ended another development would occur that would add to Phillips' excitement about the potential of oil. On December 17, 1903, at Kill Devil Hills four miles south of Kitty Hawk, North Carolina, two young Ohio bicycle shop owners would successfully fly. The wooden frame of ash and spruce tied together with waxed linen cord and covered with Pride of the West White Muslin was powered by an gasoline engine of their own creation. The distance covered by the little machine's initial flight was modest, but the event was "…the first in the history of the world in which a machine carrying a man had raised itself by its own power into the air in full flight…"[261] And, just like the "gas wagons," the Wright brothers' machine ran on gasoline.

Over the next forty-seven years, Frank Phillips would become the epitome of the oil tycoons of the early twentieth century and build one of the world's leading oil companies. Men such as Harry Sinclair, Bill Skelly, E. W. Marland, and J. P. Getty were his contemporaries. Ironically, W. W. "Bill" Keeler, the grandson of George Keeler of Nellie Johnstone No. 1 fame, would become a Phillips employee in 1929; over the next thirty-nine years he rose to become the chairman of the company. In 1949, President Truman appointed Bill Keeler as Principal Chief of the Cherokee Nation.[262]

Over sixty other tribes including the Quapaw and Cherokee spread across Indian country. Fifteen hundred statuesque, fiercely brave, and warlike Osage joined the tribes in 1872. The final home of the Osage was a large tract of northeastern Oklahoma with grass as high as a horse rider's head. Initially, it was the lease revenues from cattlemen whose herds grew fat on the rich grasslands that provided the Osages with their first taste of prosperity.[263] The location proved to be fortunate for the Osage tribe for reasons other than the excellent grasslands. However, the presence of oil was not known at the time of their settlement. But soon derricks rose across Osage County's prairie grasslands and proceeds from the oil lease auctions poured over three hundred million dollars into the Osage Tribe coffers during the first seventy-five years of production, making them the richest Indians in the world.[264] The Osage had acquired their lands through a deed from the Cherokees and not by treaty with the government. This meant there would be no excess lands for white settlement.[265] Each Osage tribe member was allotted 658 acres. However, the mineral rights were held for the tribe as a whole. Thus, all of the oil wealth would be distributed equally among those with a share or head rights.[266] This arrangement was a significant factor in allowing the Osage to retain a great part of their wealth.

Although there had been production on the Osage reservation since 1897, it would be 1904 before the Osage would truly enter the petroleum picture with the wells at Avant.[267] And it would be in the Osage reservation that Phillips would have some of his greatest successes during the formative years of the company. Although Phillips acquired his share of the Osage oil leases through competitive bidding, he also developed a close relationship with the Osages, in particular Osage Chief Fred Lookout and his wife, Julia, of Pawhuska.[268] Such was the relationship and mutual respect between Frank Phillips and the Osages that the tribe adopted Phillips as a member on September 27, 1930, the first white person ever adopted into the tribe.[269]

Even before the Bartlesville discoveries beginning with the Nellie Johnstone No. 1, oil was being sought elsewhere in Indian Territory. By September 1895 and prior to the approval of allotments by the secretary of the interior, A. W. Abrams and others obtained mineral leases from a number of Quapaw allottees. Abrams, along with Samuel Crawford, the Quapaws' perennial lobbyist, bypassed government objections to the leases and secured congressional approval for the Quapaws to lease their allotments for agricultural and business purposes with the exception of those Quapaw who were very old, idiots, imbeciles, or insane. With passage of the legislation on June 17, 1897, the Quapaw was not only the lone tribe to achieve self-allotment, but was the only tribe that could lease their lands without government supervision.[270] But Abrams and Oklahoma Territorial governor William C. Renfrow had not waited for the passage of the government's approval. In 1896, they formed the Quapaw Oil and Gas Company and drilled unsuccessfully for oil on the leases they had secured on substantial areas of the reservation. But it was on one of Abrams own farms that another valuable resource was discovered in 1897. While digging a water well on the Maud Abrams farm due east of the family home, a heavy vein of lead ore was found.[271] Four months earlier in that same year, the newly incorporated community of Bartlesville less than seventy miles to the east would also celebrate the discovery of oil.

The discovery of lead on the Maud Abrams farm would become known as Lincolnville, but this was not the first discovery of the mineral in Indian Territory. Since 1891, a small amount of lead and zinc had been mined about four miles southeast of Lincolnville and just south of the Quapaw reservation boundary near present-day Peoria. Later Oklahoma discoveries occurred in 1904 just west of Quapaw and in 1907 at the rich Commerce field.[272] Although Oklahoma was the last member to join the tri-state lead and zinc mining boom, it would become the greatest producer of all, and the Quapaw reservation would contain the mother lode with the discovery of the Picher field in 1915.

Unlike oil, the presence of significant deposits of lead in the region had been well known for many years. Interestingly, the Quapaws were peripherally associated with those first French explorers whose actions in claiming the Louisiana Territory would result in the first commercial lead mining west of the Mississippi River. Only nine years after the Quapaws' first contact with the white man, LaSalle had met the Quapaws on the banks of the Mississippi on March 13, 1682. He would continue his journey to the Gulf of Mexico and claim all of the Louisiana territory for France's Louis XIV. A relatively short thirty-eight years later in 1720, another Frenchman by the name of M. La Motte discovered significant lead deposits along the St. Francis River in eastern Missouri near present-day Fredricktown.[273] The lead ore discoveries of 1720 were only 250 miles north of the Quapaws' home at the mouth of the Arkansas. One hundred ninety-five years later, vast reserves of this same ore would be found beneath their reservation in northeastern Oklahoma.

It was not until the acquisition of the Louisiana Territory in 1803 that lead deposits in western Missouri were first noted. For the next half-century, hunters, explorers, and miners occasionally visited the southwestern Missouri region and returned with reports of wild beauty and natural resources including evidence of lead ore and its smelting by Indians and hunters. In 1838, John Cox became the first settler in the southwest Missouri region when he built a cabin and established a post office that he called Blytheville located within the boundaries of present-day Joplin. Blytheville's name would eventually be changed to that of a Methodist minister who arrived in 1841 to minister to the spiritual needs of the area. Harris Joplin built a cabin and place of worship on a spring-fed stream that was soon called Joplin Creek by the locals.[274]

Indians, hunters, and trappers had smelted southwestern Missouri surface lead deposits in crude, makeshift fires for decades. But some historians credit William Tingle as being the first to commercially mine lead ore in 1848 in what would eventually become the Tri-State Mining District, the richest lead and zinc-producing region in the world. Here our story intersects with another and much more widely recognized discovery—the California gold fields. Johann Sutter, a failed storeowner from Berne, Switzerland, decided to build a sawmill fifty miles north of the trading post. James Marshall, a mechanic from New Jersey, erected a sawmill for Sutter on a branch of the American River in the lower Sacramento Valley about forty miles from present-day Sacramento. On January 24, 1848, Marshall discovered gold on the property and told Sutter of his find four days later. Word of the discovery spread rapidly, and the gold rush began in earnest ten months later when President Polk confirmed the discovery in his annual message to Congress on December 5, 1848. The term Forty-niner was born as thousands from all over the United States and from around the world rushed to California in hopes of striking it rich.[275]

So the story goes, Don Campbell was one of those thousands on their way to the California gold fields when he stopped at his uncle's farm near present-day Joplin. William Tingle regaled his visiting nephew with tales and Indian lore of buried treasure from waylaid Spanish caravans. Tingle also talked of other metals found in the area from which Indians made shot for their guns. Campbell, along with a Negro slave loaned to him by his uncle to carry Campbell's equipment, went exploring on his uncle's farm and found an outcropping of free lead ore. Tingle and Campbell soon had smelted two wagonloads of lead chunks. Oxen pulled the heavily laden wagons to Boonville, Missouri, where the lead was shipped to La Salle, Illinois. After Campbell received fifty dollars for his half share, he continued his journey to California, perhaps never knowing the fortune he abandoned for the lure of the western gold fields.[276] Another account states that a man living on the branch near Tingle's home discovered a small amount of lead in 1848. Tingle immediately started digging and didn't stop until he had dug up a hundred pounds of lead.[277]

Following reports of the discovery on the Tingle farm, an experienced miner by the name of David Campbell visited John Cox in 1849 and remarked on the favorable geologic indications of the presence of lead ore on Cox's property. Campbell's assessment occurred at the very same time a Negro youth owned by Cox found some heavy cubed material while digging for fishing bait along Joplin Creek. After taking his strange discovery to his owner, Cox built a fire and smelted the youth's discovery into refined lead ore.[278] Several historical publications give varying or mixed accounts of the first discovery—whether it be on the Tingle farm or on the Cox property in the Joplin Creek Valley, whether it was the Negro lad on the Tingle farm or the Negro lad working for John Cox, or whether it was Tingle's nephew Don Campbell or Cox's miner friend David Campbell.[279]

Tingle's farm, located two miles west of the John Cox property, would become known as Leadville Hollow. John Cox's property was located on Joplin Creek within the city limits of present-day Joplin.[280] Following other discoveries in 1849 and 1850, new settlers including experienced miners arrived almost weekly. As more discoveries were made, mining camps and the beginnings of future towns emerged. Smelters and other businesses needed to support the miners and mining operations soon developed.[281] These mining camps and small towns were concentrated in and around the Missouri towns of Joplin, Grandby, and Oronogo. The ore deposits ranged between surface outcroppings down to a depth of fifty to sixty feet and therefore were relatively easy to mine. The ore deposits would be found from Wentworth through Joplin to Oronogo, a distance of twenty-five miles.[282]

Mining activity was significantly curtailed during the Civil War as Confederate and then Union forces fought over and operated some of the mines as a source of lead for ammunition for their respective armies.[283] After the destruction of the fledging mining industry that began in 1848, it would be five years after the end of the Civil War before a revival of interest would occur as new discoveries were made in August 1870 along Joplin Creek. By the summer of 1871, two competing mining camps were platted as towns, Joplin surrounding the original discovery on Joplin Creek and Murphysburg a quarter of a mile to the west.[284] The magnitude of the crime and lawlessness that existed in these rival mining towns cannot be overstated. The winter of 1871–1872 was known as the Reign of Terror. Without police and with a sheriff located many miles away in Carthage, the dregs of society ruled the days and nights of the camps. Shootings, street fights, saloon fights, drunkenness, prostitution, and all manner of other crimes were perpetrated without restraint. Through the efforts of the more responsible citizens, the two camps were eventually joined under the name of Joplin on March 23, 1873. Gradually, order was established, and the civilizing elements of families, churches, and an organized society began to take hold. By 1875, the population had grown to over five thousand and included 1,620 schoolchildren taught in local schools. Joplin boasted twenty-seven grocery stores, five hotels, two banks, sixteen physicians, sixteen abstract and law offices, and dozens of dry goods stores, clothing stores, blacksmith shops, livery stables, lumberyards, and boot and shoe stores. But with seventy-five salons, Joplin remained far from tame.[285]

The fortunes of Joplin would experience cycles of growth and then decline as new discoveries in other areas of the district would temporarily attract some of its population. However, Joplin became the only true small metropolitan center in the Tri-State Mining District as a result of annexation and by becoming a supply and trade center, transportation hub, and home to many miners who worked in the surrounding mining camps. Diversification ensured Joplin's preeminence among the eighty-one mining camps that existed at various times during the mining district's history.[286] Joplin had over forty years after its official founding to consolidate its position before the district's last and greatest discoveries of lead and zinc was made in Oklahoma. Had the Picher discovery occurred much closer to the time of Joplin's birth, it is interesting to speculate on the eventual fate of the city. Could Picher or perhaps Miami have supplanted Joplin as the dominant city in the region?

The southeast Kansas segment of the mining district was growing almost simultaneously with that of its neighbors in southwestern Missouri. Cherokee and Crawford Counties were formerly part of the Cherokee Neutral lands owned by the Cherokees from 1835 until 1866. This strip of land lay in the southeastern corner of Kansas immediately north of the Quapaw reservation in the northeastern corner of Oklahoma. Small discoveries of lead were made in 1872 in Garden Township and near Baxter Springs. In April 1876, the first major lead ore deposit in Kansas was discovered by accident at the bottom of a well on the Harper farm. About a mile northwest of the Quaker colony in Short Creek, it would become known as the Bonanza mining camp. Mining progressed slowly and few rich deposits were found. In February 1877, another discovery was made a mile southeast of the Bonanza camp on the Nichols farm near Short Creek by two boys digging a hole next to the roots of a tree. The two prospectors who developed the Bonanza camp sank a shaft and hit a pocket of high-grade ore at fifteen feet. The Discovery, as the Short Creek location was called, was announced on April 2, 1877, and birthed the mining boom in southeastern Kansas. Galena lay adjacent to and immediately south of Empire City and would incorporate the following day. Following election of city officials, each city inducted their respective leaders on the same day and at the same hour. Civic pride was not confined to Missouri's Joplin-Murphysburg feud of 1871–1873. The Galena-Empire rivalry was just as intense as the rivalry between its Missouri neighbors.[287]

A block-long east-west street called Red Hot connected the south end of the north-south street of Columbus in Empire City and the west end of the east-west Main Street in Galena. Red Hot was infamous for its numerous corrupt businesses that catered to the miners who rubbed shoulders with the street's usual inhabitants of outlaws, murderers, gamblers, prostitutes, and other outcasts of society. The body of a missing miner would often be found at the bottom of a nearby mineshaft following a visit to one of the Red Hot establishments such as Round Top, The Log Cabin in the Lane, or Dick Stapps.[288]

Such was the animosity between the two cities that on July 25, 1877, the Empire City Council decided to erect a stockade fence a half-mile long and eight feet high on its southern border to prevent the free interchange and flow of traffic between the towns. The wall was built under the protection of armed guards. An angry Galena City Council quietly organized a posse of fifty Galenaians and at 4:00 a.m. on August 5 attacked and burned most of the offending wall. It would take another thirty years before hostilities and jealousies between the two communities would officially end on July 9, 1907, when Empire gave up its status as a separate town and became a part of Galena.[289]

Progress in civilizing the rambunctious beginning of Galena and other wild mining camps may be directly attributed to men such as O. W. Sparks. Born in 1863, Sparks came to the Galena mining fields as a young man and soon become a millionaire several times over. To Sparks and his wife, Ida, were born three children. Ida died in 1902, and several years later Sparks married his older children's piano teacher, Miss Ambrosia Newton. In the second year of their marriage, Ambrosia was expecting a child. So proud of the pending birth, Sparks presented his wife a broach containing forty-two perfect blue diamonds suspended on an Amethyst chain. O. W. Sparks, Junior, was born but would be nicknamed Zeke. Upon graduation from Galena High School, O. W. Senior presented Zeke with a Stutz Bearcat. The Sparks occupied a mansion that covered almost a city block of Galena.[290]

But Sparks' generosity was not confined to his family. When the ore would play out in one of his mines and his workers would have to be laid off until a new operation would start up, Sparks would get the Galena merchants to extend credit to his men for food and clothes until a new mine could be established. Then, he would go and pay all of his miners' accumulated bills. This practice continued during the Depression until Sparks' death on June 30, 1932. Such was the popularity of O. W. Sparks that he was elected as a Democrat in predominately Republican Cherokee County, first as mayor of Galena and later as sheriff of the county. Eventually, he would win elections as a Kansas state representative and state senator.[291]

The first major discoveries of lead ore in the Tri-State Mining District were made in and around Joplin in 1848 and again in the early 1870s. The second major discovery was made in Kansas in the late 1870s. A third major discovery occurred just before the turn of the century and encompassed the Missouri towns of Wentworth, Prosperity, Carterville, Webb City, Oronogo, Carl Junction, and Waco. Prior to the huge Picher deposits found in Oklahoma in 1915, the region had approximately 500 mines in Missouri, 100 mines in Kansas, and a scattered handful in Oklahoma.[292]

The trials and tribulations experienced by Joplin and Galena in their early days were replayed many times throughout the Tri-State Mining District as wild mining camps struggled to put on the cloak of civilization and respectability and become towns and cities in their own right. A few made it. Most are largely forgotten with their locations marked only by piles of chat, rusting mining machinery, or the remains of a few building foundations.

In the last quarter of the nineteenth century, southwestern Missouri and southeastern Kansas became the princes of lead and zinc ore production. Not too many years after the dawn of the twentieth century, a new king would be crowned in the new state called Oklahoma. But the kingdom of lead and zinc would compete with oil to see which would produce the most millionaires.

Many a Missouri and Kansas mining man must have cast a curious if not covetous eye at Indian Territory and thought of the potential riches that lay under its grass-capped prairie surface. But government restrictions on the white man's dealings with the Indians effectively prevented exploration and mining activity through the end of the nineteenth century. A young man from Tennessee would become one of the first white men to search for subterranean riches in Indian Territory. John McNaughton had experienced a lot of life in his first twenty-four years. He had been a teamster, a bullwhacker on a wagon train, and a muleskinner on a freighting crew. In his travels to Arizona and Old Mexico, he learned several Indian languages as well as various mining methods. In September 1877, McNaughton, heading north to the Black Hills of South Dakota to do some prospecting, was accompanied by a penniless Shawnee Indian hoping to be rejoined with his tribe. The Indian told McNaughton tales of Spanish mines reputed to be located on the Peoria reservation. Intrigued, McNaughton went to Seneca, Missouri, a town that lay on the Missouri side of the Missouri-Indian Territory border a dozen miles south of the intersection of the Missouri, Kansas, and Indian Territory borders. By 1873, mining had begun four miles south of Seneca. McNaughton wasted no time in renting a buckboard and team and riding west into Indian Territory. Reaching the Spring River about five miles east of the border, he turned north toward the Peoria reservation. Upon his arrival at the supposed Spanish mines about five miles northwest of Seneca, he was amazed to see a forty-acre site with several crude shafts and tunnels that had been dug over an extended period of time by the use of primitive stone implements. Archeologists would later study the 250 to 300-foot deep shafts and conclude that the diggings were merely flint pits used by ancient Indians as a source of flint rock for the making of arrowheads. McNaughton descended the shafts by means of rope but did not find evidence of gold or any other minerals of value.[293]

Still believing that the Spanish may have mined the area, McNaughton returned to Seneca and submitted a request to the Indian Agent to be allowed to prospect in the Peoria reservation. Hiram Jones, agent in charge of the Seneca Indian Agency, denied his request. Undeterred, McNaughton found a wealthy backer in Sherman, Texas, and following a trip to Washington, D.C., obtained a special permit to prospect for minerals from the Secretary of the Interior under President U. S. Grant. McNaughton's subsequent prospecting just west of the old Spanish mines resulted in his discovery of lead, not the gold he sought. Although authorized to prospect, the Secretary of the Interior would not permit McNaughton to lease lands from the Peorias nor develop mines. Disheartened, McNaughton moved a few miles north to Baxter Springs, Kansas, and began working in a clerical position in a store owned by Peoria chief Jim Charley. He met and married Clare E. Peery, daughter of the purchasing agent for the Peoria tribe and granddaughter of Baptiste Peoria. Now as an intermarried Peoria Indian, he and Clare moved back to the reservation. In 1889, Congress allotted each member of several tribes, including the Peorias, 200 acres of land each. Having waited for over ten years, McNaughton leased several thousand acres of Peoria land for mining. The Peoria Mining Company was organized in 1891 with McNaughton as Vice President, and the first shaft was sunk northeast of Peoria. However, delays and legal problems disheartened McNaughton's Kansas investors. Not discouraged, McNaughton took options on his partners' shares, sought capital in New York, and organized The Peoria Mining and Construction Company in February 1892.[294]

Given the much larger Missouri and Kansas discoveries of the 1870s and 1880s, a lack of infrastructure necessary to support the mining industry, and legal problems surrounding the leasing of Indian lands, the Peoria discovery created little interest or excitement. The most productive period for the Peoria mines was from 1891 to 1903, but those years only produced minerals valued at an average of approximately $7,500 per year.[295] The value of lead and zinc concentrates for the entire period from 1891 through 1928 was only about $194,000, or $5,250 per year.[296]

The second discovery of lead ore in Indian Territory occurred by accident in 1897 on the Maud Abrams' farm due east of the family home. The adoption of Abner Abrams as a member of the Quapaw tribe is recounted in a previous chapter. The Abrams farm was located just southeast of present-day Quapaw, Oklahoma. The Sunnyside mine was the first shaft sunk at the Abrams discovery and would become Lincolnville. Twelve hundred miners occupied the camp by 1903. Ore was sporadically found over a two-mile area extending westward from Lincolnville. A landowner would become a millionaire from royalties because of lead and zinc found on his property while a neighbor across the road would remain in poverty because of a lack of ore deposits on his land. Most Quapaws remained poor while a few became fabulously wealthy. Unlike the Osage, the Quapaw tribe did not retain their mineral rights in the name of the tribe. The mineral rights belonged to the allotment owner.[297]

In 1905, four Miami, Oklahoma, men had pooled their money and were drilling exploration holes in search of lead and zinc deposits. With money nearly exhausted, their fifth hole struck an extremely rich vein of ore about four miles north of Miami.[298] James Robinson, Charles Harvey, and brothers George L. and Alfred Coleman would eventually become the dominating force on Indian land mining development. All but Charles Harvey were water-well drillers based in Miami, Indian Territory. In 1895, Robinson sold his drilling rig and began working in Harvey's Miami general store and later became partners with Harvey in real estate and insurance. While drilling a water well on Emma Gordon's 200-acre allotment, the

Coleman brothers discovered a substantial amount of lead and zinc in the bore hole cuttings. The brothers discussed their findings with Robinson and Harvey. The men formed a partnership and began quietly obtaining leases surrounding the Emma Gordon allotment. As a member of the Eastern Miami tribe with an allotment with the Quapaws, she was not able to enter into a valid lease until August 1907. Drilling began on the partners' various leases in late 1905, and it would be almost a year before word leaked out about the men's discovery. By early 1907, the men had been unable to secure financing from the local banks nor interest Joplin's mining men in the discovery because of their disbelief in the richness of the ore. Lacking capital to continue drilling on the leased lands, the four men decided to sublease some of their land now held in their Miami Royalty Company. Mining activity began with the sinking of the first shaft in 1907.[299]

Robinson, Harvey, and the Coleman brothers had tried to obtain financing from the Bank of Miami. However, the bank considered loans to develop mining lands in 1905 and 1906 to be too risky. The four partners had waited two years to lease the Emma Gordon land but found themselves unable to compete with others seeking to lease her 200-acre allotment. The prized lease went to the owners of the Bank of Miami, and less than a month later the men would lease the valuable land to Amos D. Hatton for a ten-year period. The Emma Gordon mine would become one of the richest producers of lead and zinc in the area. Robinson was humiliated by the bank's failure to provide financing. That the same owners of the bank acquired the coveted Emma Gordon lease must have rubbed salt into his wounded pride. A biography of Robinson relates that he said, "Some day I will own that bank." And he did.[300]

The town of Hattonville was established near the Miami Mining Camp located about four miles north of the City of Miami. The town's name was changed to Geneva (Hatton's daughter's name), and then back to Hattonville again. Although not able to obtain the Emma Gordon lease, the four owners of the Miami Royalty Company owned leases covering eight thousand acres of the richest mining land in the area. On October 14, 1913, the assets of the Miami Royalty Company were transferred to a trust called the Commerce Mining and Royalty Company including the town site of Hattonville. Hattonville was subsequently platted as Commerce and would later become the hometown of future baseball star Mickey Mantle.

It is ironic that the discoveries of immense quantities of two natural resources would occur in the same year (1897) on the reservations of two Indian tribes resettled in close proximity to each other in northeastern Oklahoma. Over the next one hundred years the fortunes of each tribe and the lands they occupied would stand in stark contrast to each other. One tribe with great wealth and a sense of community is centered among relatively prosperous cities and towns while the few remaining members of the other tribe live amid pervasive poverty on reservation lands dotted with mountains of mine tailings, polluted streams, and rusting mining equipment.

But in the decade of statehood and the decade to follow, hope sprang eternal among the wildcatters and prospectors. *Never mind the dry hole or mineshaft without pay dirt, the next one will make us rich!* And sometimes the dream came true.

Boomtown!

"But they that will be rich fall into temptation and a snare, and into many foolish and hurtful lusts, which drown men in destruction and perdition."[301]

The western portion of the 5.5-mile by 16-mile Quapaw reservation that lay in the northeastern corner of Indian Territory was for the most part a flat, treeless grassland. Little could be heard except the rustle of grass as a rider urged his horse across the prairie, the bawling of a lost calf searching for its mother in the tall blue-stem, or the gentle, mechanical singing of sickle mowers cutting the rich grasses for binding and shipping amidst the protest of an occasional prairie chicken. A few trees in the western half could be found crowding the banks of a creek here or there, but most would be found along the beautiful Spring River flowing out of Kansas and meandering southward near the Oklahoma-Missouri border among the gentle hills of the eastern half of the reservation.[302] But the tranquil countryside would soon become a clamorous mass of men and machines gouging holes in the land and building mountains of chat, the leftovers after extraction of valuable lead and zinc.

Even during the first decade of the twentieth century, the peaceful meadows and rippling streams of the Quapaw reservation were being increasingly disturbed by the shuttle of men and machines traveling between the Kansas and Missouri mining fields and those in Indian Territory—the Peoria field to the southeast, Miami fields to the southwest, and Lincolnville within the reservation's own borders. Travel was difficult at best and nearly impossible when rains turned the few dirt roads and virgin prairie into a rutted quagmire. Horses pulling the heavy iron and steel drilling rigs were no match for the dirt roads that became little more than mud tracks after a rain.

It was at such a time as this in 1913 when Richard (Dick) Blosser, a mining man and contract driller, was moving one of his drill rigs back to Baxter Springs, Kansas. Blosser had been drilling near Hattonville, Oklahoma, for O. H. Picher and his son, O. S. Picher, of the Picher Lead and Smelting Company of Joplin. The Pichers decided the Hattonville site on which Blosser had been drilling was not worth mining. The Pichers directed Blosser to move his machine back to Baxter Springs, Kansas, where they had other leases to be drilled. Recent rains had turned the prairie hay field over which Blosser had been moving his machine into a swamp. The drill rig became mired in the rain-soaked field.[303] During this time the Picher Mining Company and Eagle Lead Company were merging.[304] Since the Blosser drill rig couldn't be moved until the ground was either frozen or dried out, the owners of the new Eagle Picher Lead Company eventually instructed Blosser in late 1913 to drill where the rig was stuck. But the heavy machine had to be gotten up out of the mud before drilling could commence. Three-inch by twelve-inch planks eighteen feet long were located and acquired from a railroad company in Baxter Springs. These planks and others bought from a lumberyard were hauled by wagon to the drill site, and the drill tools were retrieved from where they were left near Hattonville. With jacks and considerable effort, Blosser and his men were able to raise the steam engine-powered drilling machine a little at a time to a point where drilling was ready to start.[305] This first hole yielded exceptionally rich pay dirt and was located three-quarters of a mile south of the Kansas-Oklahoma border and one and one-half blocks west of Connell Avenue in present-day Picher.[306] Five more drill holes were made in close proximity to the first by moving the drill machine over the planks.[307]

The company took great pains to keep their discovery quiet. All of the drill hole cuttings were buried, the sludge pond destroyed, and even the driller's daily logbook was taken. It would be several weeks before the discovery was announced. The announcement came in August 1914 that the Eagle Picher Company was completing exploratory drilling on one of the largest ore discoveries ever known.[308] By the summer of 1915, the company had drilled approximately thirty holes that radiated in all directions from the original drilling site. It was at this time the company sank the first shaft on the Crawfish lease.[309]

The above account was based on Howard Blosser's *Prairie Jack-pot*. Howard Blosser was Dick Blosser's son and an eyewitness of many of the events described in his book. Howard Blosser's account is an invaluable source of information based on his personal experience. However, the 1913 dates appear to be inaccurate with regard to the occurrence of those events. Other sources give slightly different versions and dates for the Picher discovery. One account states the drill machine was going to Hattonville as opposed to returning to Baxter Springs as stated in Blosser's account. The source also dates the first borehole as being made on March 15, 1915.[310]

Arrell Gibson's excellent and scholarly *Wilderness Bonanza* states that the drill rig was stuck in the mud near Tar Creek in August 1914. While waiting for help the driller sank a hole that indicated a rich vein of ore lay below the surface. Following their driller's report, the Picher Lead Company had quickly and quietly obtained leases on 2,700 acres surrounding the initial find.[311] The timing of the discovery in August 1914 makes more sense in that the Blosser dates of 1913 are almost two years before the first shafts were sunk in the Picher field. Such long periods between a discovery and the sinking of mineshafts to recover the newly discovered ore are not typical of other discoveries in the Tri-State. The discrepancy is understandable given the fact that Mr. Blosser recorded his boyhood memories sixty years later.

Whatever the details and date of the drilling of that first borehole, all the world would soon know of the mining camp that would become Picher, Oklahoma. Although the first shaft was not sunk until mid-1915, the magnitude of the ore deposits produced by the Picher field is obvious when one considers that 90% of the Tri-State Mining District's production would come from this field after 1915.[312] By the end of 1915, there would be more than 160 mines and mills within a five-mile radius of Picher, the rough boomtown that was to become the world's greatest lead and zinc discovery. From an economic standpoint, the huge discovery of ore could not have been timelier with the on-set of World War I. Zinc prices rose from $40 per ton to as much as $135 per ton.[313] The highest average annual price for zinc concentrates in the Miami-Picher District between 1908 and 1930 was $74.80 per short ton and occurred in 1916. This was more than double the district's average annual price for 1914.[314]

Such was the growth of Picher that the town created in 1915 on the empty, grassy prairie had grown to be the largest in Ottawa County with a population of 8,172 according to the December 1918 special census report. Miami was second at 6,898. In addition to Miami, other incorporated towns in the county were Cardin, Commerce, Afton, Quapaw, Fairland, North Miami, Wyandotte, Peoria, and Narcissa. Ottawa County grew from 15,713 to 43,692 by the end of 1918, a growth of 278% in eight years. The county newspapers prominently noted the lack of a "foreign element" in the county's population comprised of 93.7% white and 6.7% Indian. The 6.7% amounted to 2,756 for all of the tribes located within the county's borders.[31]

Panoramic view of early day Picher, Oklahoma (Photo courtesy of Baxter Heritage Center & Museum, Baxter Springs, Kansas)

All of the elements of a classic boomtown culture were present in Picher—fortunes made overnight; an influx of thousands of men and a few women hoping to at least improve their lot in life if not become one of the few to strike it rich; a dearth of supplies and services to meet the daily needs of food, shelter, and other basic necessities; and an absence of the restraining influence of family and law enforcement.

The lands leased from the Quapaws were under the direction of A. E. Bendelari, representative of the Eagle Picher Company in Picher. Because of the company's substantial control of the land in and around Picher, Bendelari became the ultimate authority when disputes arose in the mining camps.[316] The original town site of Picher was surveyed and platted on eighty acres under Bendelari's direction, and the first lot was sold in early 1915. As the land belonged to the Quapaw Indians, land sales had to be approved by the U.S. Department of Labor at that time. The first lots were leased for thirty-day periods.[317]

Food and lodging became the immediate concern of all new arrivals to Picher. The town site was soon dotted with a jumble of tents, tiny one-room houses made from blasting powder boxes, and numerous other shacks of various materials and descriptions.[318] One enterprising fellow from Joplin owned a piano firm. Seeing an opportunity to use his large piano crates, he shipped them to Picher, lined them up on Main Street, placed a mattress in each, and charged seventy-five cents per night's lodging in his piano crate hotel. Other houses were on wheels or skids pulled to the mining sites. Still other crude shelters were built by cementing together syrup cans, whiskey bottles, and rocks.[319] Mr. and Mrs. W. C. Jones established the first grocery as soon as the survey was completed and a lot-lease secured. Their tent-store was pitched at 103 North Main. Burt Luther bought a lot in early 1915 at Second and Connell, and he is believed to have been the first resident on Connell Street in the newly platted Picher. He also established a grocery, and both the Joneses and Luther were still in business as late as 1927.[320]

Life in early Picher was not for the faint of heart. The hardships were numerous. The dirt streets and sidewalks became swamps with citizens sinking to their ankles in mud following a rain. But it would not be long before the clouds of choking dust would again rule the atmosphere and co-exist with the multitude of sounds emanating from the shrieking mine whistles, the chug-chug of the steam and gasoline engines pulling ore cans from the bowels of the earth, and the constant ebb and flow of men and machines between the various mining camps. To these sights and sounds were added the smells of sweating bodies, horses, and livestock, and the ever-present one and two-hole outhouses. But dust, noise, and stench were far more bearable than the harsh winters spent in unheated tents or packing crates and scorching summers under a cloudless sky on a treeless prairie. Such was life in the first years of Picher that it soon earned the reputation of being one of the wildest of the mining camps in the central United States.[321]

Sickness and disease must be added to the inherent dangers of early mining practices that threatened life and limb of miners and other residents. A 1915 pink eye epidemic was blamed on impure water. Two years later quick action lead to the prevention of all but three deaths in a 1917–18 smallpox outbreak. But the infamous influenza epidemic that swept the nation and world in the fall of 1918 would claim the lives of 300 Picher citizens.[322]

Mining accidents and disease were not the only threats. The promise and presence of money and wealth always attract the less desirable elements of any society. Gamblers, confidence men, swindlers, thieves, prostitutes, sellers of illicit hard liquor, and other purveyors of vice that preyed upon the hardworking miner and other citizens were present from the beginnings of Picher. Apart from the efforts of Bendelari as head of Eagle Picher, Picher's first unofficial governing body was the Commercial Club organized March 3, 1917. Out of this body came the Law and Order League, a vigilance committee organized in the fall of 1917, which served to bring a measure of accountability to the lawbreakers of the rowdy community. These two organizations would prepare the groundwork for the incorporation of Picher in April 1918. Dr. D. L. Connell would become the head the board of trustees and the town's first chief executive. Doc Curtis was hired as the town's first marshal and would quickly take action to clean up the lawless community.[323]

Some central plains rivers are described as being a mile wide and an inch deep. In many respects this is an apt description of early-day Picher—a mile wide in its population growth and mining activity but only an inch deep in the civilizing influences that change residents into citizens, houses into homes, and a mining camp into a respectable town. But these civilizing influences, however small, would begin almost immediately after Picher's birth.

The first electric lights were installed in the newly organized non-denominational Union Church that also served as the first public school in 1916.[324] But the Union Church was not large enough to house the two thousand school-aged children that resided in and around Picher by 1917. As Indian land was not taxable, the Commercial Club raised $5,000 in donations to build the first school buildings on College Street. Initial construction of roads, bridges, and the purchase of the first fire-fighting equipment were also paid for by donated money. To continue building the necessary infrastructure and pay for town services, a gross production tax was imposed on each ton of ore sold by the mine operators.[325]

The first Chamber of Commerce organized in 1919 lasted only a few months. Its successor, the Advertising Club, lasted three years. A second Chamber was established in July 1926. To fill the void left between the Advertising Club's demise in 1922 and the new Chamber in 1926, a Kiwanis Club was organized in Picher in 1924 and was responsible for many civic improvements. By the end of 1927 and the first dozen years of Picher's existence, deep waterwells were drilled, two and one-half miles of concrete pavement were poured, and connecting highways to surrounding towns were established. Maps of the period show a maze of rail and electric lines for both freight and passenger traffic connecting Picher with numerous cities and towns from as far away as Carthage, Missouri, to the northeast and Miami, Oklahoma, to the southwest.[326]

Picher would have been a boomtown without the occurrence of

World War I. However, Picher's boom was significantly enhanced by the events occurring in 1914 a half a world away. The assassination of Austrian Archduke Franz Ferdinand on June 28, 1914, was the spark that led to the war and the creation of many boomtowns of a different kind that stretched along the network of trenches crossing Europe from the North Sea to the border of Switzerland near the Mediterranean and in other parts of the world.[327] While the excited drillers examined the lead rich drill cuttings from that first Picher bore hole under a blazing Oklahoma sky in August 1914, these small European villages and towns on what would become known as the Western Front began to reverberate with the explosion of millions of artillery shells. The barrages would last for over four years following the beginning of the war in that same month of August 1914.

World War I was the first machine-dominated war. Tanks, submarines, airplanes, and poison gas were innovations that made their appearance in World War I. But the great killers of the war would be machine guns and artillery. A machine gun could fire from 450 to 600 rounds per minute, the equal of fire from forty to eighty riflemen. These high-volume of fire weapons were made possible by smokeless powder that meant less bore fouling. Other weaponry innovations included faster operating bolt-action magazine rifles, steel-jacketed bullets, and solid-brass cartridge cases.[328]

The war, with these new machines, would consume vast amounts of lead and zinc. As newspaper headlines reported the events unfolding in Europe, it took little imagination for the Tri-State Mining District entrepreneur and miner alike to become excited about the money to be made in mining lead and zinc. Added to the war news were the exciting reports about the new strike of vast proportions just north and east of Miami and Commerce.

The tremendous increase in lead demand can be seen when the Franco-German War of 1870–71 is compared to the opening weeks of World War I in the summer of 1914. A rifleman in the Franco-German War was provided 200 rounds of ammunition, and the average infantryman would expend only fifty-six rounds in six months of war. By contrast, the World War I rifleman would receive

280 rounds, which were used up in the first weeks of the war. In the 1870–71 conflict, the Germans fired an average of 199 shells per artillery piece. In the first weeks of World War I, the German used all of the 1,000 rounds made available for each gun.[329] The French began the war with 1,300 rounds of ammunition for their guns, barely a three-week supply.[330]

A great usage of shells occurred in the pre-battle shelling of enemy forces. In the eight days preceding the battle of the Somme in July 1916, 1,732,873 artillery rounds were fired at an average rate of fire of 9,000 shells per hour, or 150 shells per minute.[331] Along the nine and one-half mile front at the Battle of Messines stretched 2,571 guns, one gun about every twenty-one feet.[332] In the eight days preceding the July 1917 battle, 3,258,000 shells were fired at an average rate of fire of 17,000 shells per hour, or 282 shells per minute. The havoc wrecked by the millions of exploding artillery shells transformed the landscape into a sea of mud, craters, and tangled wire, all devoid of vegetation, through which the foot soldiers struggled and artillery guns could not follow.[333]

The development of the quick-firing gun allowing artillery barrages such as the ones at the Somme, and Messines began with the French in 1897. By 1914, many but not all nations had adopted and equipped their armies with the new artillery.[334] The main artillery projectile at the beginning of the war in 1914 was the shrapnel shell, a steel tubular projectile filled with lead balls. As the projectile curved downward in its trajectory, a fuse would ignite a charge in the base of the shell, and the lead balls would be blown forward from the shell at a height of about thirty feet above the ground. Any troops below the cone-shaped spray of lead balls would be injured or killed. A sixty-pound shrapnel shell contained 990 lead balls and weighed twenty-eight pounds. A fifteen-pound shell contained 200 balls and weighed almost six pounds.[335] Prior to World War I, military strategists anticipated a war of maneuver punctuated with periodic battles. But World War I quickly settled into a relatively static setting with opposing forces facing each other in a series of roughly parallel trenches and bunkers separated by a no man's land of tangled barbed wire. In such an environment in which earthen barriers, trenches, and bunkers generally protected troops, the shrapnel shell was not as effective as the high explosive shells fired by the larger guns. As the war progressed, the larger guns firing high explosive shells claimed a much greater role. At the beginning of the war, France had only 389 heavies out of a total of 5,108 artillery pieces. By war's end, the French had almost as many heavy guns (5,600) as they had field pieces (6,000).[336]

The enormity of the artillery war is almost incomprehensible. During the entire war, the British fired 170 million rounds with an average shell weight of almost sixty pounds and collectively weighing five million tons. In September 1917, almost a million rounds were fired in a twenty-four hour period. During the Third Battle of Ypres, 4,283,550 rounds were fired in a two-week period. To the trench soldier, a light barrage was considered to be a half-dozen shells landing close by about every ten minutes. In a heavy bombardment, twenty to thirty shells would land in a company's sector each minute. Generally, there would be a mixture of three large, high explosive howitzer shells and one shrapnel shell. The barrage would stop for five minutes and then begin again. As the day progressed, the intervals between barrages grew shorter until there was almost a continuous series of explosions.[337] Artillery was the greatest killer of the war with some estimates attributing to it 70% of all battlefield deaths.[338]

The direct costs of the war are estimated to have been $180 billion along with $150 billion in indirect costs. Sixty-eight million soldiers fought in the war of which ten million died (including two million who died from disease and malnutrition). Twenty-one million were injured. Almost one out of every three soldiers in the war were injured and one in seven were killed. Civilian deaths attributable to the war totaled 6.6 million.[339] By any standard of comparison, the percentage of troops killed in World War I were staggering for the major combatants: British Empire, 10.6%, Germany, 16.1%, France, 16.1%, Russia, 14.2%, and 15.4% for Austria-Hungary. Even more astounding are the percentages of forces killed or wounded: British

Empire, 35.2%; Germany, 64.9%; France, 73.2%; Russia, 76.3%; and Austria-Hungary, 90%. When one considers that these figures include all troops mobilized, not just those on the front lines, the changes of a trench soldier being killed or wounded are even greater than these percentages suggest. The magnitude of the carnage is revealed when one considers the number of war-related deaths as a fraction of the combatants' entire populations: British Empire, 2.4%; Germany, 3.8%; France, 3.6%; Russia, 2.4%; and Austria-Hungary, 5.2%. Some of the smaller combatants suffered much greater percentages of loss of their populations: Bulgaria, 8.3%; Serbia, 17.6%; and the Ottoman Empire (Turkey), 10.1%.[340]

The fact that the price per ton of zinc concentrate rose faster than that of lead during the early years of the war gives an indication of its importance in the war effort. The greatest need for zinc was its usage as a zinc-copper alloy to make the millions of brass shell casings for rifle, machine gun bullets, and artillery and mortar shell casings. Sag paste (sag was gas spelled backwards) was an ointment made from zinc and applied to the skin by soldiers to protect themselves from the burns caused by poison gas. Zinc dust was burned to create smoke screens to protect troop positions and hide ships at sea from submarines. Zinc had many other important uses in conducting the war effort including the manufacturing of rubber tires.[341]

The impact of the war's demand for lead and zinc on Picher cannot be overestimated. And after 1915, the Picher field provided 80% of the district's production. Considering the escalating price for lead and zinc during the war, the timing of the Picher discovery could have not been better. The Tri-State District was the world's largest supplier of lead and zinc until about 1945, producing one-tenth of the United States' lead and one-half of its zinc.[342]

Lead and zinc are sold as concentrates. Only a small percentage of the ore produced contains lead and zinc concentrates. The milling process to be discussed later separates the concentrates from the remainder of the ore (mine tailings) that is discarded. The price per ton[343] of lead concentrates in the Miami-Picher District was $47.19 in 1914, $54.60 in 1915, $83.94 in 1916, $100.77 in 1917, and $88.54 in

1918, the last year of the war. Between 1919 and 1930, the price per ton of lead concentrates ranged between $64.68 and $117.13 with the exception of 1921 when the post-war average price per ton dropped to $52.48. The district's zinc sulfide concentrate prices per ton were $32.66 in 1914, $66.48 in 1915, $74.80 in 1916, $67.44 in 1917, and $50.55 in 1918. Between 1919 and 1930, zinc concentrate prices per ton ranged between $30.09 and $50.64 with the exception of 1921 when the post-war average price per ton declined to $22.83.[344]

Along with the escalating prices for lead concentrate during the war, production in the Miami-Picher District increased significantly, in a large part because of the amount of ore flowing from the Picher mines. District lead concentrate production totals were: 9,058 tons in 1915, 15,306 in 1916, 32,765 in 1917, and 69,862 in 1918. Lead concentrate production in the district peaked at 103,359 tons in 1925 before gradually declining to 39,000 tons in 1930. Zinc concentrate production increased even more dramatically than lead concentrates: 28,280 tons in 1915, 54,935 tons in 1916, 159,656 tons in 1917, and 300,702 tons in 1918. As with lead, zinc concentrate production peaked in 1925 at 549,211 tons gradually declining to 268,171 tons in 1930. To give perspective to the Miami-Picher production, lead concentrates amounted to 1.01% of the ore mined and milled in the peak production year of 1925. In other words, every ton of lead concentrates produced would require approximately one hundred tons of ore be extracted from the mines and hauled to the surface. Zinc concentrates amounted to 5.13% of the ore mined and milled in that same year. Therefore it would take approximately twenty tons of ore to produce one ton of zinc concentrates. In total, 10.6 million tons of ore were mined and milled in 1925 to obtain 653,000 tons of lead and zinc concentrates.[345]

The Picher field began emptying its bowels of lead and zinc in 1915. In the next four years, much of its production would be used to rain terror and destruction on the European continent a half a world away. The shattered landscape would take decades to heal. Ironically, the scars left by the mining operations that produced the lead and zinc would far outlast the destruction caused to Europe in World War I.

Brother Lead and Cousin Zinc

"And God said, Let the waters under the heaven be gathered together unto one place, and let the dry land appear: and it was so."[346]

As one writer expressed it, "Picher was the end of the rainbow, the fairy spot where the Almighty was said to have broken his apron string when he was sowing riches and to have spilled the greatest part of his treasure of lead and zinc there."[347]

A line from an old Frank Sinatra song about love and marriage says, "You can't have one without the other." This is an apt description of lead and zinc in the Tri-State mining region. Rarely was one discovered without the presence of the other. Not only were the two minerals produced from the same mines, the Tri-State district's mills were adapted to process both minerals in the same operation. But irrespective of the lyrics of Mr. Sinatra's song, a better understanding of each ore is had when their histories and uses are separately examined.

The first known uses of lead were for small statues, ornaments, and pottery glazes by the ancient Egyptians 6,000 years ago. The Chinese used lead coins as long as 3,000 years ago, and in the ninth century just south of present-day Baghdad, lead sheeting was used to line the floors of the Hanging Gardens of Babylon. Just as the Babylonians used lead sheeting to hold soil and keep in the moisture, the Greeks used lead as a protective covering for their boats and roofs of their houses. The famous Roman aqueduct system would not have been possible without the use of lead. Their plumbing and water supply system used pipes made from rectangular sheets of lead bent around wooden poles and soldered at the joined edges.[348] So prevalent was the association between the Romans and their lead water pipes that the modern word "plumber" comes from "plumbum," the Latin word for lead. From that word also came lead's chemical symbol Pb in the periodic table of elements.[349]

The Romans created a lead acetate by boiling sour wine in lead cooking pots. The chemical reaction caused by the vinegar in the wine and the lead in the pots produced lead acetate, or "sugar of lead." These white crystals were called "sapa" and used to sweeten food, and the Romans consumed large amounts of the lead-based sweetener. The Romans consumed the sweetener even though they were aware of its poisonous properties and harmful effects on their health. Some scholars believe that the physical and mental impairment caused by the lead content was one of the main reasons for the fall of the empire.[350]

By the middle ages it was common to see the roofs of churches and cathedrals made from lead, including a cathedral built in the thirteenth century at Chartres, France, that lasted over 600 years. Lead was used in the making of pewter ornaments and items such as plates and mugs, and the earliest firearms used bullets made from lead. Lead type was used with the invention of the printing press in Germany in the 1400s.[351]

In each of these ancient civilizations and societies, it was the alchemists who were the scientists and metalurgists of their time. The alchemists' knowledge and secretive experiments in laboratories filled with strange contraptions amid the hiss of flames beneath bubbling cauldrons were every bit as mysterious to the king and peasant alike as Einstein's theory of relativity and the complexities of the double helix of the DNA chain are to us in the twenty-first century. Over the course of centuries the alchemists came to believe that of the seven metals, gold was the most perfect substance of all. Silver, mercury, iron, lead, copper, and tin were nature's failures. [352]

The birth of alchemy was shrouded in mysticism as alchemy was regarded as a sacred art. This association with the spiritual resulted in the belief that the various metals were inherently good or evil. Gold was identified with goodness. Lead was considered the oldest of the metals and was representative of sin and evil.[353] However much we may be amused by the ideas, assumptions, and hypotheses of these ancient alchemists, their work would eventually result in a body of valuable scientific knowledge with regard to the properties of metals and the creation of various alloys that made possible new inventions and the advancement of civilization.

Although early day alchemists ascribed the production of metals in the earth to the influence of celestial bodies, modern scientists have proposed three theories as to the origin of lead and zinc ores in the Tri-State region. Two of the theories propose that the ores were present in small amounts over widely distributed sedimentary rocks. The concentration of the ores into rich deposits was accomplished over an unknown number of millennia by cold waters, either descending ground waters or ascending artesian waters. The third theory suggests that ores were deposited by ascending thermal waters originating in the magmatic region far below the surface of the earth.[354]

Under the first of the two water disposition theories, evaporation of shallow seas in the Tri-State region helped concentrate solutions of minerals that eventually saturated the rock crevices and faults of the limestone beds that underlay the area. The second theory holds that ascending artesian waters picked up the widely dispersed minerals and deposited the solutions in reservoirs present in the faults of the rock. This is called primary enrichment and is complimented by the first theory of descending surface and ground waters, which is termed secondary enrichment. Crystallization of the lead and zinc fused together the rock fragments and faults. Proponents of the third theory believe that ore deposits originated in rocks near molten magma far below the earth's surface. Under this theory, heat from molten magma caused gases and vapors of the magma to penetrate the pores of the mineral material in the rocks and resulted in a chemical reaction that forced a concentration of ores including lead and zinc. Heated artesian waters carried the molten solution upward to the reservoirs beneath the surface of the Tri-State region.[355]

Regardless of which theory to which one subscribes, scientists are in general agreement that the Ozark Uplift provided the necessary repository for the various ore deposits. The Uplift is the product of considerable geologic disturbances over the ages. The region is known for its many caves, but these represent only a small fraction of the region's numerous cavities, openings, fissures, caverns, and channels created by the flexing and shattering of the earth beneath the surface.[356] Into this porous layer just below the surface of the Tri-State region were deposited one of the earth's richest concentrations of lead and zinc. And the richest part of the Ozark Uplift was centered on Picher, Oklahoma.

The shapes of the lead ore bodies under the earth's surface generally fall within four types: flat runs, vertical runs, ore pockets, and blanket veins. Flat runs are broad, tubular ore bodies that range from ten to thirty feet in height and may be as little as ten to twenty feet or up to 500 to 1,000 feet long. Vertical runs range as much as 150 feet high with widths ranging from ten to fifteen feet. Ore pockets tend to be small and circular in shape with deposits being very fractured or fragmented.[357] Blanket vein ore formations, as the name would indicate, are thin, horizontal sheets of ore deposited over a wide area. The thickness of the deposit range from as little as a tiny fraction of an inch to as much as twenty feet. One famous blanket vein ore bed was located around and through Webb City, Missouri, and was a continuous sheet that extended for twelve miles. Flat and vertical runs were most prevalent in the Oklahoma fields. Blanket vein deposits were found mostly around the Webb City area.[358]

Several forms of lead are found in the Tri-State Mining District. Galena or lead sulfide contains 87% lead with the remainder being sulfur. Lead sulfide is the most prevalent type of lead found in the district and is noted for its purity and ease of smelting. Other secondary lead ores contain lesser amounts of lead due to weathering and are mined nearer to the surface. Lead carbonate is called drybone by the miners. Pyromorphite, or green lead as it may be called, is lead phosphate, and the third secondary lead ore is anglesite, or lead sulfate.[359]

If for no other reason than its exceptionally heavy weight, early settlers in the Tri-State region recognized the presence of lead when evidenced by surface outcroppings. Lead's relatively heavy weight comes from its density. It is eleven times more dense than water and one and one-half times as dense as most other common metals.[360]

Lead also has a distinctive appearance. Galena appears as crystals in the form of cubes that vary in size from minuscule to some as large as a foot. The bright lead-gray crystals with a metallic luster may appear either as individual cubes or cubes with corners truncated by octahedron faces. These crystals would be found as individual crystals, groups of crystals, or as fine grains scattered through the rock.[361] When exposed to oxygen, galena turns a dull, bluish-gray.

Lead was used from ancient times for water pipes, metallic ornaments, pottery glazes, bullets, lining in hulls of boats and ships, coverings on roofs, lead type, paint, burial vault liners, and a host of other applications. Lead usage beginning in the early part of the twentieth century was to undergo a remarkable transformation. When the automobile's engine crank went the way of the buggy whip, the lead necessary for making rechargeable lead-acid batteries dramatically increased the demand for lead. Presently, 250 million automobile batteries are produced each year with the substantial majority of the lead used in those batteries coming from lead salvaged from used lead-acid batteries.[362] Other modern uses of lead include the making of ammunition, creation of an alloy by mixing tin and lead to make solder used to join metals together, making lead glass for camera lenses and other optical instruments, and making paints containing lead compounds for protection of iron and steel from corrosion. However, because of its poisonous properties, lead paint was banned in 1978 from use in homes and other buildings where frequent human contact is possible. Unleaded gasoline became available in 1975, and lead was completely banned as an engine anti-knock gasoline additive in 1995.[363]

As any fan of Superman knows, lead is so dense that even Superman's X-ray eyes cannot penetrate it. However, the value of lead's protection for Superman of the comics, radio, television, and movies has real-life applications in medicine and industry. Because of lead's ability to absorb potentially harmful radiation, medical technicians wear lead-lined aprons when performing X-rays on patients. Radiation from nuclear reactors is contained by use of lead shielding. Leaded glass is used to provide protection from radiation when the operator must have a clear view, both in X-ray rooms and nuclear power plants.[364]

China produced 30% of the world's mined lead in 2004. The United States, Mexico, Australia, Peru, and China accounted for 80% of the world's lead mine production in that same year. The world's identified lead resources are estimated to be 1.5 billion tons.[365] If recoverable, the lead resources would last almost 500 years at the present annual rate of consumption.

A significant shortage of lead would occur if mine production was the world's only source of lead. Recovery of scrap lead accounts for twice as much lead production as from mining in the United States. In recent years 97% of the lead in used lead-acid batteries has been recycled. This amount is significant when one considers that the production of lead-acid batteries consumes 83% of the reported lead consumption in the United States. Ammunition, casting, radiation shielding and other sheeting, pipes and plumbing products, building construction uses, solder, and oxides for glass account for 11% of U.S. consumption. The remaining 6% of U.S. consumption is used in ballasts and counter weights, bronze and brass, foil, various metal alloys, type metal, wire, and miscellaneous uses.[366]

For centuries cousin zinc was the Rodney Dangerfield of metals. Not only did it not receive respect, zinc was not even recognized as a member of the metals family until about 1400. The seven metals of antiquity were discovered from approximately 6,000 BC (gold) to around 750 BC (mercury). Another 2,000 years would pass before another four metals were discovered in the thirteenth and fourteenth centuries and added to the original seven—arsenic, antimony, bismuth, and zinc.[367] Zinc's value to ancient civilization was found in its ability to form brass when alloyed with copper. The Bible lists Tubal-cain, a seventh generation descendent from Adam (in the lineage of Cain), as an artisan in brass and iron.[368] Brass relics made from a mixture of copper and zinc and dated from 1400 BC to 1000 BC have been found in ancient Palestine.[369] These ancients may not have understood the scientific process of making brass, but they knew how to use zinc ore to make the valuable alloy. Brass may contain 55 to 95% copper when mixed with zinc, and evidence of the earliest know use of brass date to 2,500 years ago. The Romans used brass extensively for coins, decorative pieces, and pots.[370] Zinc was smelted both in India and China as early as AD 1000.[371] By 1400, zinc was recognized as a new metal in India where both zinc metal and zinc oxide were refined from the twelfth to sixteenth centuries.[372] Credit for the discovery of pure metallic zinc in the West is generally given to Andreas Marggraf; in 1746 he heated zinc carbonate (calamine) with charcoal.[373] Almost simultaneously, William Champion established a zinc smelter in Bristol, England.[374]

Zinc was used mostly as a building material until a variety of uses were discovered during the industrial revolution of the nineteenth century and subsequent technological developments of the early twentieth century. In the early lead mining fields of the Tri-State region, the value of zinc was not recognized and generally considered a waste product. This view probably resulted from a combination of ignorance and a lack of smelting facilities. As one legend goes, an unsuspecting Joplin man sold his solid zinc wall in 1861 for $1,750 replacing it with ordinary stone. The foolishness of the seller's bargain soon was widely known and established a healthy respect for the value of zinc.[375]

Perhaps no other metal has found such a wide variety of uses as that of zinc. There are six major categories of zinc usage: anti-corrosion coatings for steel and iron; die cast metal components; brass; construction materials; pharmaceuticals and cosmetics; and nutritional supplements for humans, animals, and plants.[376]

The author's first experience with zinc occurred as a teenager in the summer between my freshman and sophomore years at college. Working the night shift in a galvanizing department of a metal fabrication plant was a dirty and dangerous job. Twenty-foot-long steel trays with precariously balanced loads of angle iron and steel stacked three or four feet high were hoisted by an overhead crane and then submerged in large acid vats to remove rust and foreign matter in preparation for galvanization. Following a quick rinse in a similar-sized vat of water, the acid-cleaned metal would subsequently be dipped piece by piece in a molten zinc bath to give the metal an anti-corrosion coating. These molten zinc baths—rectangular heated pits built into the floor of the plant measuring approximately four feet wide and twenty to thirty feet long with a depth of five feet or more—would be filled with molten zinc. This was the hot-dip method of applying a protective zinc coating. The hot-dip method may be performed in a batch process by submerging individual iron or steel items in the molten zinc as described above or in a continuous process where a ribbon of steel is fed through the molten zinc prior to further fabrication. The hot-dip method is used when a thicker layer of zinc is required when the iron or steel item will be subject to a particularly corrosive environment such as steel girders used to build bridges in or near salt water. The pool of zinc is diminished as the galvanized steel and iron carry away a coating of zinc. The pit must be replenished by use of an overhead crane to deposit solid blocks of zinc weighing hundreds of pounds into the pool of molten zinc. These zinc blocks soon melt in the 860 degree Fahrenheit cauldron.

A second method of applying the zinc coating to iron and steel is called electrogalvanizing, an electroplating process. This process applies a thinner coat that will be painted on items not needing the thick coating of zinc as applied by the hot-dip method. Electrogalvanizing has substantially replaced the hot-dip method. Roofing materials, automotive body parts, and consumer appliances are examples of things usually electrogalvanized.

Coating steel and iron with a thin coat of zinc is one way to afford protection from corrosion. Another method of protecting exposed steel or iron structures is by using sacrificial anodes, commonly called cathodic protection. Sacrificial zinc rods or plates (anodes) are placed adjacent to and electrically connected to the structure to be protected from corrosion. This causes the anode to corrode or sacrifice itself over time as opposed to the structure protected. By regulating the current, the anodes become the negative electrode. Examples of structures that are cathodically protected from corrosion include underground petroleum storage tanks, bridges, and oil pipelines.

There are few if any metals that match zinc for its wide variety of applications. It is used in the making of rubber tires, paint, fertilizer, dyes, roofing materials, and various electronic components. As mentioned in the previous chapter, zinc was used in World War I to protect troops from burns caused by poison gas. Called sag paste, soldiers would smear the zinc oxide on exposed skin or as an ointment for treating burns. Smoke screens were created by burning zinc dust during the war to hide ships at sea or soldiers on the battlefield from enemy gunners. Zinc is used as a deodorant (zinc chloride), to create the luminescent pigments for the glow in the dark hands of a clock or watch (zinc sulfide), to treat skin rashes (calamine), in vitamins and mineral supplements, for sunburn protection, and as a remedy for the common cold (zinc glycine in lozenge form).[377]

Zinc is an important nutrient for humans, animals, and plants. Zinc deficiency may lead to growth retardation or stunting, immune system disorders, loss of appetite, and loss of hair to name only a few. The World Health Organization has estimated that a third of the world's population has insufficient intake of zinc. Zinc deficiency in soil may result in significant reduction of crop yields.[378]

The main zinc ores are sphalerite (zinc sulfide), smithsonite (zinc carbonate), calamine (zinc silicate), and hydro-zincite. The most important ore in the Tri-State mining region was sphalerite, which contains 63% metallic zinc and 33% sulfur. Jack, rosin jack, black jack, or blende were the miners' names for sphalerite. Smithsonite, called drybone by the miners,[379] was named for James Smithson, an English chemist and mineralogist. In 1802 he proved that zinc carbonate was a true carbonate as opposed to zinc oxide. A wealthy man, Smithson left his considerable estate to the United States of America to found the Smithsonian Institution in Washington, D. C., with the purpose of promoting the increase and diffusion of knowledge.[380] It would be 1838 before the 105 sacks of gold sovereigns valued at a half-million dollars ($50 million by today's standards) would be transported to the United States.[381] In November 1846, in spite of years of political wrangling of Congress and through the tireless efforts of former president John Quincy Adams, the red sandstone building that sits on the National Mall today was commissioned to be designed and built to house the institution that would begin the fulfillment of the terms of James Smithson's will.[382] The building that would become known as the Castle opened in 1855 and was the beginning of one of the world's greatest educational, scientific, cultural, and historical institutions.[383]

Since lead and zinc ore deposits are usually found together, it is not surprising that the major producing countries for lead also produce the most zinc ore. China led with 22% of the world's zinc production in 2004. Ninety percent of the United States' production came from Alaska. Identified world zinc resources total 1.9 billion tons and will last over 200 years at the current rate of production. In the United States, 400,000 tons of zinc was recovered from waste and scrap in 2004. Of the total consumption of zinc in the United States in 2003, 52% was used for galvanization, 22% for zinc-base alloys, 17% for brass and bronze, and the remaining 8% for all other purposes.[384]

Into the Pits

"Hear me speedily, O Lord: hide not thy face from me, lest I be like unto them that go down into the pit."[385]

It is difficult to imagine the feelings, thoughts, and fears of a young man of sixteen as he prepared to descend for the first time into a mine that would be his home for a third of every day except Sundays and an occasional Saturday. He had heard the old miners talk of the mines, using jargon that seemed foreign to those not familiar with mining and the camps that surround them. These men were not large as he originally imagined. Many were small, but all appeared lean and work hardened with shoulders and arms well developed from shoveling rock and ore into ore cans eight hours a day. The men seemed older than their years; the haggard look and persistent cough from breathing too much mine dust had hidden their true ages. Although most of the men were only in their twenties and thirties, their manner and appearance were of men much older. It was what the mines had done to them. It was as though they left a little bit of themselves each time they emerged from the mines. The young miner promised himself he would not stay that long.

Men appeared from every direction as the mine whistle's distinctive call beckoned them to another day underground. The line of men snaked toward the pull derrick above the shaft. His apprehension grew as the number of men between him and the black hole in the ground grew smaller and smaller.

He had traveled far from the isolated Kansas farm on which he had grown up. He wished he could return, but there was no work there to be had. As a state, Oklahoma was less than a decade old. He wanted to make his mark in this newest member of the Union, but he would need money to do it. Perhaps he could save enough to buy some land and run cattle or farm. Making four or five dollars a day in the mines was real money. But there were expenses too. Seventy-five cents to rent a mattress in a piano crate, and food was expensive. *Can*

I do it? he wondered. Could he make it as a miner? And, more importantly, would he survive? Only yesterday when he asked for work at the mining office, they had just brought out the crushed body of a miner unfortunate enough to be standing beneath a slab of rock that had broken loose from the ceiling. Only now did he realize that he was the dead man's replacement.

The line of men dwindled as three or four men at the head of the line stepped into the hip pocket high bucket measuring thirty-three inches across that was hooked to a cable controlled by the hoisterman sitting in his perch above their heads. With lunch bucket in one hand and a death grip with the other on the cable or bail of the bucket, the miners disappeared as the bucket plunged toward the bottom of the shaft. As quickly as they disembarked, the empty can flew back to the surface with each round trip lasting less than a minute.

Only two men were ahead of him now. He would go down in the next bucket. All too quickly the bucket made its round trip and was at the surface again. One of the men steadied the bucket as the young man's foot slipped. The man grinned and said, "It's a long way down. Better take the bucket." No sooner had the third man stepped into the bucket than the bottom seemed to drop away as the can plunged downward. The young man chanced a look upward as they sank into the well of darkness. The square of light from the mineshaft opening was rapidly growing smaller. In seconds the young man's stomach and knees felt the deceleration of their fall as the bucket approached the bottom of the shaft. The young man scrambled out but stumbled as his eyes adjusted to the darkness. Lighting the little lamps on the front of their hats and starting down a dark passage, one of the men laughed and called to the young man, "Grab a shovel and follow us, you're a miner now!"

———

When people think of mining, they visualize deep holes, tunnels burrowed through the earth, and experiences such as those of our imaginary young friend above. Mining is a simple process. Dig a hole and take out the ore. If the ore is near the surface, it can be scratched out with a stick. But the ore is rarely at or near the surface, so the holes must get deeper.

Early methods of determining where to dig for ore ranged from the semi-scientific to superstition and folklore, such as "witching" with a willow branch or consulting fortune tellers. Prospecting was not only wildcat drilling or sinking a shaft in an unexplored area but included developmental drilling to expand areas of known ore reserves.[386]

After determining where to drill or sink a shaft, the mining process encompasses three major phases. The first phase is the extraction of ore from the ground; milling or separation of the desired ore from dirt, rock, and debris is the second phase; and the third phase is refining the ore into metal. Each of these functions evolved over time. In the Tri-State Mining District, this evolution occurred in three major periods: the early period prior to 1900, the intermediate period from 1900 to 1930, and the modern period from 1930 through 1955.[387]

Early lead and zinc mine in the Tri-State (Photo courtesy of Baxter Heritage Center & Museum, Baxter Springs, Kansas)

Mine derrick—men on derrick (Photo courtesy of Baxter Heritage Center & Museum, Baxter Springs, Kansas)

Tub hooker and ore bucket—Admiralty No. 3 Mine (Photo courtesy of Baxter Heritage Center & Museum, Baxter Springs, Kansas)

Mule being lowered into mine shaft (Photo courtesy of Baxter Heritage Center & Museum, Baxter Springs, Kansas)

Miners working in water—St. Louis No. 4 Mine—1941 (Photo courtesy of Baxter Heritage Center & Museum, Baxter Springs, Kansas)

Power monkeys—Dynamite in the mines—Red-Skin Mining Company (Photo courtesy of Baxter Heritage Center & Museum, Baxter Springs, Kansas)

Indiana Mine cribbing (wooden walls of mine shaft), water pump, and pipes leading to the surface (Photo courtesy of Baxter Heritage Center & Museum, Baxter Springs, Kansas)

*Two-man outhouse in a mine (Photo courtesy of Baxter Heritage
Center & Museum, Baxter Springs, Kansas)*

*Roof trimming on an eighty-foot ladder (Photo courtesy of Baxter
Heritage Center & Museum, Baxter Springs, Kansas)*

*Pillars left in mine to support roof (Photo courtesy of Dobson Museum &
Memorial Center, Miami, Oklahoma)*

*Hoisterman (in back) and screen apes breaking boulders into pieces
small enough to fall through the rails—Rialto No. 1 Mine (Photo
courtesy of Baxter Heritage Center & Museum, Baxter Springs,
Kansas)*

In the early period, the Tri-State miners were almost universally owner-operators involved in all three aspects of mining—excavation, milling, and crude efforts at refining. Due to the shallowness of deposits and lack of competition from larger, mechanized mining operations, a man could begin mining in the Tri-State region with little financial resources; in fact, the Tri-State District became known as the "poor man's camp."[388] Recall the work of William Tingle and nephew Don Campbell (pp.124-125) when they discovered an outcropping of lead on Tingle's farm near present-day Joplin. The men excavated the lead, separated the ore from the dirt and rocks, and refined the ore into lead chunks. In the days before 1900, the miner and his partner would dig a four by four feet or four by five feet pit with a pick and shovel and explosives when necessary. As the pit became deeper, the miner would descend in a bucket tied to a rope attached to a windlass operated by his partner. The windlass is a horizontal post or barrel supported at each end by two vertical posts, one located on each side of the shaft. A crank at one end would allow a rope to be wound around the rotating post or barrel and thereby raise or lower the bucket. The same bucket ridden by the miner to the bottom of the pit was also used to haul up dirt and rock. Many of these pre-1900 prospecting pits would be abandoned when the water table was reached, usually at a depth of about fifty feet. If the quantity of lead and zinc ore found above the water level were not sufficient to justify further mining, the pit would be abandoned.[389] If ore samples in the side of the shaft indicated enough ore to justify development of the mine, a tunnel or "drift" would be started in the side of the shaft.

With the ore available from the shallow mines exhausted by 1900, it was increasingly necessary to go deeper to find ore. However, deeper prospecting shafts were expensive and also lacked a suitable means of removing water from the mineshafts. Due to these limitations, the churn drill was becoming the tool of choice for deeper exploration. Between 1890 and 1900, the churn drills were stationary platforms and drill towers built of lumber over an area to be drilled. The drill tool was approximately six inches in diameter and several feet long and was attached to the end of a rope or cable. The other end was wrapped around a drum used to wind or unwind the rope or cable, which would raise or lower the drill tool. Animal or steam power would cause the drum to wind up the rope or cable and raise the drill tool. When released by the driller, the drill tool, weighing 1,800 pounds, would plunge to the bottom of the hole and cut into the earth or rock deeper than the last drop. A brake drum allowed the driller to rapidly raise and drop the drill tool a few inches many times each minute. Cuttings from the tip of the drill tool were periodically examined to determine the presence of ore. Each time a drill hole was completed, the drill tower and platform would have to be dismantled and rebuilt at the location of the next hole to be punched into the earth. The efficiency and speed with which an area could be prospected was greatly improved with the introduction of mobile churn drills adapted from those used in the nearby oil and coalfields. These raise and drop machines were known as cable tool rigs in the oil fields. Generally, churn drills could reach depths of 150 to 250 feet at an average rate of penetration of ten to fifteen feet per day.[390]

Around 1900, mining in the Tri-State began evolving into big business. Companies well capitalized by faraway investors began supplanting the two-man partnerships digging fifty-foot shafts. Professional mining engineers, consolidation of small mine holdings, modern equipment, technological innovations, separation of labor from ownership, labor specialization, and improved transportation infrastructure occurred in the intermediate period of Tri-State development between 1900 and 1930. These years saw the growth of organized labor in the mining fields and a new emphasis on mine safety and miner health. Not only did the excavation phase of the mining process change, milling evolved too. Early milling occurred on site right next to the mineshaft. Due to larger volumes of ore and the improved efficiencies of large scale milling operations, mined ore was transported to larger, centralized mills located near a group of mines.

Mineshafts measuring five by seven feet or six by six feet became the standard. In the Oklahoma portion of the Tri-State Mining District, mineshafts ranged from about one hundred feet deep in the eastern part to 400 feet in the western section.[391] It was essential that the mineshaft be vertical or the ascending or descending ore cans would hit the sides of the shaft and damage the cribbing or dislodge rocks. The wooden walls of the mineshaft, called cribbing, were constructed of two by six-inch timbers affixed to the sides of the shaft. Some shafts were only partially cribbed while others were completely walled with timbers from the top opening, or collar, down to the platform at the bottom of the shaft. Cribbing lessened the danger of falling rocks and the hooking or catching of the free-swinging bucket on the shaft walls as it was raised or lowered. Once the bottom of the shaft was reached and a drift was to be started, the shaft would be extended an additional five to twelve feet to create a hole to collect water to be removed by a sump pump. Water removal lines, cables, and conduits were fastened to the cribbed walls as they rose to the surface. All men, equipment, and ore cans rode the cable controlled by the hoisterman. The area surrounding the platform at the bottom of the shaft was enlarged to allow maneuvering of men and ore cans waiting for transport to the surface and returning empty cans.[392]

Once a mineshaft had reached the ore level, drifts were cut radially into the walls of the shaft. The drifts would lengthen as rock and ore were removed from the end of the drift and hauled back to the shaft in cans. If ore deposits were extensive, drifts were pushed farther and farther from the shaft and additional shafts would be dug, generally when the end of the drift was over 300 feet from the first shaft. In the early days of mining in the Tri-State, cans of rock and ore were slid along the drift floors on heavy oak-planked runways kept wet with water. The ore cans could not exceed 150 pounds; any can heavier could not be pushed to the mineshaft and hoisted to the surface. The oak planks were replaced by wooden rails upon which small flat-bedded four-wheeled carts were used to transport the ore cans to the shaft. By 1900, lightweight steel rails replaced the wooden tracks.[393]

Once the shaft was complete, drifts or tunnels were cut into the sides of the enlarged area around the platform at the bottom of the shaft. Most drifts expanded into large underground rooms with pillars of un-mined rock and ore left in place to support the roof of the cavern, usually about 15% of the ore body. A pull drift connected the shaft with the room or drift containing the ore body or connected one ore body with another. The location and amount of ore determined the shape, width, and height of the room, or stope as it was called. Typically, pillars supporting the roof would be twenty to fifty feet in diameter and thirty to one hundred feet apart. For ore to be loaded into cans for transport back to the shaft and then to the surface, the rock and ore had to be broken loose from the wall or working face. The minimum height of a working face was six feet and could be as high as one hundred feet. To provide broken rock and ore for loading, a drill man would drill holes into the working face, and the powder man would load the holes with dynamite. The holes could be drilled as deep as twenty feet into the working face. The drill holes started at 2.75 inches and tapered down to 1.5 inches as the holes became deeper. Blasting occurred at the end of the day so that the dust would settle and allow inspection of the roof for loose rock or slabs of stone before arrival of the miners the next morning.[394]

Roof trimmers performed one of the most dangerous and well-paid jobs in the mines. A roof trimmer used a long metal rod to pry loose any rocks or slabs from the mine's ceiling. Since most ceilings were not reachable from the floor of the stope, a ladder had to be used. The roof trimmer was the aerial artist or trapeze act in a mine's circus of performers. With some ceilings over one hundred feet high, he often worked far above the floor of the mine. Mine ladders, made of selected spruce, were capable of being extended to a height of ninety-five feet. After extending the ladder to a length equal to the height of the ceiling, four or more guy ropes or wires would be attached and the ladder raised. Each guy wire was manned by one or two miners and be equally spaced around the ladder. The roof trimmer climbed the ladder with his ten to fifteen-foot pry bar, a one-half to three-quarter-inch steel spear with a point on one end and a bent chisel point on the other end to form a pry heel. At the top of the ladder, he poked and pried any loose stones or slabs, careful not to cause the rocks to fall on him or those manning the guy wires below. As one area would be cleared of loose rocks, the men holding the guy wires would slowly pull the top of the ladder in an arc to another location. This speeded the process of examining the roof, as the roof trimmer did not have to climb down or have the ladder dismantled and moved to a new location. If a loose rock or slab proved to be stubborn and could not be barred down, the roof trimmer would load a charge of dynamite to blow it down after the room was cleared of miners. Given the unpredictability of the size and shape of loose rocks, the roof trimmer had to be a man of courage and good judgment. Not only his life but also the lives of those working in the mines depended on his skill and thoroughness.

Miners would arrive at the beginning of a shift to find tons of broken rock and ore lying next to the working face that had been blasted loose the previous evening. Prior to the blasting, wooden boards called a "lay out" were laid on the drift floor next to the working face so that the blasted rock and ore would fall on the boards. This allowed a shoveler's shovel to more easily slide under the broken rock and ore.[395]

Shovelers, sometimes called muckers, used a No. 2 scoop shovel to load the ore cans that measured thirty-three inches high by thirty-three inches in diameter. These round heavy metal buckets could hold as much as 1,650 pounds and had a two-inch diameter bail attached to each side. Shovelers were paid by the can, so they would occasionally load a "windy," a can loaded to leave as much void space as possible yet appear as a full can.[396] It would be 1940 before a mechanical shoveler could be developed that could operate more cheaply than the human shoveler.[397]

A shoveler usually loaded twenty tons of ore per eight-hour shift.[398] Cans varied in capacity from 1,000 pounds to 1,650 pounds of ore. A miner's pay varied depending on the size of the can and the number of cans loaded. A miner would have to load forty cans at 1,000 pounds each to equal twenty tons per shift but only twenty-five cans that held 1,650 pounds. Pay scales also varied depending on the type of miner and whether he worked in the boom years of the mid 1920s or the Great Depression years of the 1930s. One source states that miners earned between six and seven dollars per day.[399] One young miner working his summers between school years during the Depression reported earning $3.75 a day as a lead shoveler.[400] Another source states that the cost of loading ore was fifteen to twenty cents per ton, which reflects a daily pay rate of three to four dollars per day for the average shoveler. The same source also stated that a good shoveler could load seventy-five to ninety cans per shift, which is a remarkable feat even if one assumes the cans were of the smaller 1,000 pound capacity.[401] Such a shoveler would have to load thirty-seven to forty-five tons per shift and would earn between $5.50 and $9.00 depending on the tons loaded and whether he was paid the fifteen cent or twenty cent rate. But the back-breaking hard work of an average mine shoveler becomes obvious when one considers that the miner must load a thousand pounds of rock every twelve minutes during an eight hour shift in order to fill forty cans with twenty tons of rock and ore.

Louis Hile, born during the spring of 1915, lived in and around Picher almost all of his life. His father worked in the Peoria mines as a young boy and five years in the Picher mines in the late teens and early 1920s as a muleskinner making about $2.50 to $3.00 per day. Mr. Hile grew up during the Picher boom years, and he worked in and around the mines as a machine repairman for many years. He recalls shovelers earning ten cent per can loading about forty cans per shift, making about four dollars.[402]

To keep track of how many cans he loaded, a miner would place a paddle with his identification number in each can he filled. When the can arrived at the shaft to be hoisted, the tub hooker would remove the paddle and mark the tally board to give the miner credit. At the end of the day the shift boss would total the number of cans filled by the miner. The grand total for all miners sending cans to the shaft would be checked against the hoist operator's tally at the top of the shaft.[403]

The cans being filled sat on small four-wheeled flat cars, one can per car, that rolled on the lightweight track laid next to the broken ore and rock at the working face of the stope. Each stope had three to six miners working, and each miner shoveled ore from about five feet on either side of his can resting on the flat car. When filled, the flat car with the loaded can was pushed to a siding called a "lay by." As enough full cans accumulated in the lay by, the flat cars were hooked together to form a small train and pulled by a mule to a lay by at bottom of the shaft. Generally, the mules pulled five or six loaded flat cars.[404] If the muleskinner hooked a seventh to the small train, the mule would know it and refuse to budge. No amount of cursing or prodding would change the mule's mind.

The mine mules spent most of their lives underground. Compared to their topside relatives, the life of a mine mule was considered good. The pulling of six flat cars with cans loaded with ore was not too difficult for the beasts. They were well fed and cared for and were generally treated as pets by most of the miners. Hay barns, feedlots, fresh running water, and floored stables covered with straw were maintained underground. The temperature in the mines was constant, not too cold and not too hot. The miners' affection for their four-legged co-workers was evident by the once-a-month custom of hauling green tree branches to the mineshaft and dropping them in for the mules to eat.[405]

Every miner of the Tri-State region in the early 1900s would have at least one story to tell about the mine mules. One such story was about Peanuts, a brown mule with a reddish shine to his hide. The miners had taught Peanuts to chew tobacco. Peanuts developed such a liking for tobacco that each morning when the miners arrived, he would poke his nose into their pockets to find his daily ration of tobacco. If he didn't find any, the miner risked having his pocket ripped off. The absence of tobacco would lead to much braying and kicking followed by sulking and a refusal to work.[406]

Toby was reputed to be the last of the mules to leave the lead and zinc mines. He achieved a measure of fame as a tourist attraction at the bottom of the 220-foot deep Van Pool Mine two miles south of Picher. Toby remained in the mines long after machinery had replaced the mule for hauling ore to the shaft. But Toby continued to work hauling ore in drifts too narrow for the machines. After spending most of his life underground, Toby's last years were spent above ground enjoying his celebrity and retirement. He died at the age of thirty in 1976.[407]

Except for two occasions, mine mules were relatively well behaved and docile apart from a few eccentricities. The exceptions involved removing mules from the mines and putting them into the mines. For the mine mule, it was perhaps the preparation for the move as well as the journey to the unknown that caused them to become less than cooperative. One writer remembers witnessing such an event as a young boy. The braying beast could be heard from deep within the shaft as he progressed to the surface suspended from the cable used to lift ore cans. The mule's front and back legs were bound in a special leather harness from which only the mule's head and rump showed. Once on the ground and the last strap loosened, the excited mule jerked the rope tied around his neck from the miner's hand. Bucking, kicking, and snorting, the mule charged into a nearby blacksmith shop and knocked over the forge, causing its smokestack to fall on the roof. Having demolished much of the interior of the blacksmith shop, the mule ran out the back door and across the prairie. The formerly docile replacement mule waiting to go down into the mine was now having second thoughts and making his protestations known. The young lad, his father, and the gathered miners all laughed and enjoyed the spectacle—all except the surprised blacksmith and his helper.[408]

After the mules pulled their small train of loaded ore cans to the lay by at the bottom of the shaft, the mule handler, called a muleskinner, hooked the mule to a train of empty cans for return to the shovelers. Small electric and gasoline locomotives eventually replaced the mules and pulled fifteen to twenty loaded ore cans to the mine shaft. If the working face of the ore deposit was extended to over 300 feet from the mineshaft, it became more economically feasible to sink a new shaft than haul the ore longer distances as the cost of transporting the ore was greater than the cost of a new shaft.[409]

The tub hooker worked at the bottom of the mineshaft. He positioned loaded ore cans to be hoisted and retrieved the empty cans returned after being dumped into a hopper at the surface of the mineshaft. The bumper pushed loaded cans from the lay by to the tub hooker at the shaft platform called a dog house. The bumper also pushed away empty cans returned from the surface to a lay by to await return to the shovelers. As a can was hoisted to the surface, the tub hooker would position another flat car containing a loaded can at the center of the shaft. When the empty can returned to the bottom of the shaft, the hoisterman would brake the cable, slowing the descent of the can the last few feet above the waiting loaded can. The tub hooker then swung the empty can over the loaded can to the flat car waiting just to the side of the loaded can. As soon as slack appeared in the cable, the tub hooker quickly unhooked the empty can and snapped the hook on the loaded can. Immediately the loaded can flew to the surface, and the tub hooker would position another loaded can for its trip up the shaft. It was important that the can be properly positioned in the exact center of the shaft; a can lifted from any other position would have a tendency to swing as it rose to the surface. A swinging can would hit the sides of the shaft, causing rock to fall or damage the cribbing. The hoisterman was positioned at the top of the shaft so that he could watch the cans to see if a rock fell from a can or was dislodged from the shaft wall. In such instances, the hoisterman would quickly hit the "skidoo-bell" or light to warn those at the bottom to move away from the shaft opening.[410] Raising and lowering a can took only twenty-eight seconds from the time it was hooked, hoisted to the surface, dumped, returned to the platform at the bottom of the shaft, and unhooked.[411] This is rather a remarkable feat considering that a 250-foot shaft required the can to travel 500 feet. Allowing for eight seconds of the twenty-eight second round trip to dump the can's contents, the ascending and descending can would reach speeds of fifteen to twenty miles per hour or more. Eight hundred cans were hoisted in a typical eight-hour shift, and a record 900 cans in one shift has been recorded.[412] Eight hundred cans hoisted per shift equates to one can every thirty-six, seconds including swinging the empty cans over and unhooking them and hooking up a loaded can. If the cans averaged 1,000 pounds of ore each, 400 tons of ore would be pulled from the mine in a shift. If the cans were the larger 1,650-pound buckets, 800 cans would yield 660 tons per shift.

There were many other jobs in the mines apart from the shovelers, muleskinners, hoistermen, tub hookers, bumpers, roof trimmers, drill or machine men, and powder men (sometimes called powder monkeys). The ground boss was the underground mine foreman. His assistants were the cokey herder, or straw boss, and time keeper. A bruno man was a miner who cleaned up small amounts of ore not scooped up by the shovelers. The drill man or machine man's helper was called a dummy and carried drill bits and drills for his boss.[413] Much of the mine terminology evolved and grew with mechanization that occurred during and after the 1930s and included many jobs performed above ground and in the milling operations.

Americans have always had a quiet respect for hard-working men and women who do the difficult and dangerous jobs that are required for the people of a nation to function as a civilized society. These men and women include soldiers and sailors, policemen and firefighters, and steel workers and miners. Books and songs have been written and movies made to describe and immortalize their lives. One such song written about life in the Muhlenberg County, Kentucky, coalmines would become one of the biggest and best-loved ballads of the twentieth century by telling of the hardships and struggles of a miner. Merle Travis wrote "Sixteen Tons" in August 1946. Recorded by Tennessee Ernie Ford in 1955 as the B-side of a single, the song topped every major record chart in the United States within a month of release. Two million copies were sold in less than two months, becoming the most successful single ever recorded.[414] What chord was struck in the American public's collective psyche that so resonated as to create the song's immediate, huge, and enduring popularity? Those answers and their social implications are probably best left to the psychologists and sociologists. But a layman's guess is that people who had no knowledge of or interest in a coal miner's life or work could still identify with his life's struggles. Working hard, feeling your body's vitality and health ebb away a little each day, and struggling to pay bills and make ends meet were common ground with many Americans. In the quarter of a century between the early 1930s and the song's phenomenal popularity in the mid 1950s, Americans had survived the Great Depression, a world war, conflict in Korea, and threats of annihilation by nuclear war. Merle Travis's miner personified the average American's feelings.

And no group could better identify with the hardships of Travis's coal miner than the lead and zinc miners of northeastern Oklahoma. Irrespective of any differences between coal mining and the mining of lead and zinc, it is not hard to imagine the occasional smile on a Tri-State miner's face as he thought of his twenty tons a day in comparison to Tennessee Ernie's sixteen tons.

Mountains of Chaff

"Ye shall conceive chaff, ye shall bring forth stubble: your breath, as fire, shall devour you."[415]

The hoisterman was the link between the dark, dank world of the miner below and the dusty, raucous clamor of mining and milling operations at the sun baked or frozen surface, depending on the season of the year. If the roof trimmers were the aerial or trapeze artists of the mining circus, the hoistermen were the ringmasters. He and the tub hooker were the only two men involved with every cubic foot of ore produced by the mines, the skinny neck of the hour glass through which all ore flowed upward from the mines. Hoistermen were artists whose deft touch and acute sense of timing were critical to the successful mining operation. Befitting his status, he sat on his throne, elevated above the comings and goings of the commoners in the mines below. He operated his hoist with the skill and finesse of an orchestra conductor. Without him the mining symphony would become a cacophonous burst of confusion followed by silence.

The bottom end of the hoisting operation was discussed in the last chapter. At the top end, the muscle to lift the nearly one ton of ore and metal can was located at the top of the derrick frame about forty to sixty feet above the shaft entrance at ground level. The electric motor could spin the thirty-six inch drum or wheel to reel in its payload or send an empty can back down the shaft at cable speeds between eighteen and twenty-three miles per hour. Loaded cans were lifted at maximum speed. The hoisterman was positioned so that he could look down the shaft and watch the rising or falling bucket. When the bottom of the can was raised to the hoisterman's eye level, he stopped it and caused a counterbalanced door to close below the loaded can held by the cable. The door acted as a chute to channel the ore to a nearby bin. A ring fastened to the bottom of the bucket was caught by a suspended tail hook. The hoisterman lowered the cable hooked to the bail causing the can to tip and empty its contents into the chute. The bucket was raised again, the tail hook disengaged, the counterbalanced door opened, and the empty bucket lowered once again to the bottom of the shaft. The five-eighths inch cable had a life expectancy of about 40,000 tons.[416] With each can and contents weighing nearly a ton and 800 lifts per shift, the cable would have to be replaced every fifty working days.

As previously described, ore and rock from the loaded cans were channeled into the large hopper or bin in or adjacent to the derrick. But first the ore and rock had to pass through a series of parallel steel rails called grizzly bars spaced about four inches apart at the top of the hopper bin.[417] If the rock and ore were small enough, it would pass on through to the bin. If the rocks and boulders were too large, they would be caught by the grizzly bars. A screen ape's job was to break large rocks into small ones, somewhat like the work of convicts in less-enlightened times. Those that couldn't be broken would be tossed out into a pile next to the derrick. Most observers would agree that the convict had the better job of the two once their respective working conditions were compared. Screen apes derived their name from the orange close-fitting rubber-framed "glasses" they wore to protect their eyes from fragments while busting rocks with a sledgehammer. A circle of fine screen was affixed where the glass portion of the glasses would normally have been. The screen-covered orange glasses gave the men the appearance of having apes' faces. The screen ape's job was noisy, dirty, and dangerous. They balanced themselves on a board laid across the grizzly bars. With a bucket of ore emptied about every thirty to forty-five seconds and sliding down the metal counterbalanced door onto the steel rails and into the hopper below, there was an incredible amount of noise and clouds of dust. It is easy to understand why most miners did not aspire to be screen apes.

We have followed the ore from the working face of the stope to the shaft and up the shaft to the hopper. But the lead and zinc are still considered crude ores at this point. The subsequent milling process changed the lead and zinc crude ore into concentrates by reducing it to a uniform size, removing impurities and dirt, and separating the lead and zinc from the rock to which it may be bonded. Once the lead and zinc was reduced to concentrate form, it was ready for sale and smelting by a host of buyers who transform the concentrates into a variety of products and uses.[418]

Compared to life in the mines, mundane would be the best word to describe the milling operations. There would be no songs such as "Sixteen Tons" written to romanticize the men and their work in milling the ore extracted from the mines.

The following discussion of milling operations is by no means thorough or complete. This discussion, intended to give only a general understanding of the milling process, is necessary in order to give a complete picture of life in a mining camp. Variations in techniques, methods, and sophistication of the milling process depends on the particular years described in the Tri-State mining era from 1850 to 1950, the size of the mill, and the technology available at a particular time.

Milling is simple: keep the lead and zinc and discard everything else. Although a simple concept, the process of accomplishing this task is complex and difficult. Early milling efforts were crude and therefore wasteful in that a significant percentage of the lead and zinc content of the crude ore was thrown out with the dirt, impurities, and rock. In later years as milling techniques were improved, much of the earlier waste material was milled again to recover the discarded lead and zinc.

Free ore was lead and zinc unattached to other substances such as dirt, rock, and other minerals. By 1908, free ore had been substantially depleted in the Tri-State District, and 60% of zinc and 90% of lead were mixed with rock.[419] Milling was a continually evolving process because of advances in technology and an ever-increasing and economically driven effort to reduce the loss of ore being sent to the waste piles.

Prior to 1850, a trapper or hunter would find outcroppings of lead and merely wash away the dirt before melting the lead for bullets or bartering for other goods. Zinc ore was discarded in the Tri-

State region prior to 1874 because of its low price and difficulties transporting it to distant markets. By the mid-1870s, crude milling efforts involved breaking rock away from the ore with small hand picks. Other ore mixed with rock that could not be separated with a hand pick was crushed with a hammer. The miner would put the crushed rock and ore into the inclined sluice, a wooden box ten feet long, two feet wide, and six inches deep. The miner would continually spread the material into the stream of water coming from the upper end of the sluice. The lighter rock and other material would wash or float out at the lower end of the sluice leaving the heavier ore at the bottom.[420]

The modern milling process generally involved a series of steps: crushing, screening, jigging, tabling, and flotation of the ore. The first step was to reduce the size of the ore by crushing and occurred after ore was moved from the derrick hopper to the mill by surface tramways pulled by gasoline or steam engines. Eight to twelve tramcars would be loaded at the derrick hopper and then pulled a short distance, ranging from a few hundred feet but always less than a mile unless the ore was being transported to a centralized mill serving several mines. Upon arrival at the mill, the tramcars were pushed up an incline and the ore dumped into a mill hopper. Some mills would have subsurface hoppers over which tramcars with automatic drop bottoms would be rolled in order to drop their contents.[421] Ore is fed from the mill hopper to crushing equipment. The jaw crusher performed the initial crushing with jaws ranging from sixteen to thirty-six inches in diameter. Rock and ore from the jaw crusher had to be smaller than two to four inches in diameter before moving to secondary crushing by large breaker rolls. These large metal rollers had diameters ranging from thirty-six to forty-eight inches. Ore from the jaw crushers would be fed between the two long parallel rotating breaker rolls and reduced to pieces measuring less than a half inch in diameter.[422] As water was mixed with the rock during the crushing processes, the ore had to be deslimed or have a portion of the water removed before being fed to the jigs. The fine sands and particles that overflowed from the desliming boxes would be tabled while the courser particles would be fed to the roughing jig.[423]

Jigging and tabling, relying on the principle of specific gravity, was the basis for extracting a great majority of the lead and zinc mined in the Tri-State District. Zinc blende was one and a half times heavier than flint gravel, and lead was three times heavier. Jigging began in the district about 1870. These crude jigging devices allowed ore placed in a screened bottomed box to be jiggled inside a tank of water. The sloshing action of the water in the ore box caused the heavier ores to settle to the bottom while the lighter rock and debris would rise to the top and be skimmed off and discarded.[424]

Although considerably more sophisticated, the jigging operations of the Picher mining field in 1915 and thereafter operated essentially on the same principle as those of the late 1800s. The Tri-State District saw a great variety of jigs with many types of configurations. Generally, most roughing jigs of the era contained a series of six or seven cells. Ore would be fed to the jig cell, and a stream of water would wash the ore through the remaining series of cells, each measuring approximately thirty-six to forty-eight inches across. The heavy lead concentrates would settle in the first cell. Larger pieces of zinc blende would settle in the second cell. Smaller pieces of zinc would collect in each successive cell with the residue from the last cell being transported to the waste pile.[425]

To combat excessive losses of valuable ore, sand jigs or tables were installed. The fine sand and ore from the roughing jig was mixed with water to make a slime that was fed to a wide, slightly inclined table with a series of bars or riffles placed across it. As the slime flowed over the bars or riffles, the heavier ores settled along the ridges created by the bars or riffles and could be retrieved for sale.[426]

Flotation was a process used to further capture the valuable ores from sand that had escaped other milling processes. The ore sand mixed with water became a muddy slime that was whipped by gyrators causing air bubbles to form in the mixture. The mineral rich slime would be mixed with a chemical reagent such as pine oil for lead and copper sulfate for zinc blende. The fine grains of lead or zinc, depending on the reagent used, would cling to the air bubbles and float to the surface where they would be skimmed off and retained for sale.[427]

There were approximately 150 milling plants operating in the Miami-Picher district in 1926. Many of the mills were either old mills dismantled in the Joplin mining fields and brought to Picher or new mills of the same design. Both proved highly inefficient. One 1919 study reflected that the average Picher mill was only 63% efficient with many mills falling below 50%. In other words, almost half or more of the zinc content of the milled ore was not recovered but sent to the tailings or waste pile. Between 1925 and the end of the first six months of 1931, the amount of zinc concentrates recovered from a ton of ore ranged between 5.4% and 6%. Lead recovery averaged about 0.9%. Although the amount of lead yielded by the ore was significantly smaller than zinc, the efficiency of recovery was in excess of 90% because the extraction of lead presented fewer difficulties.[428]

As mining production began to decline after reaching its peak in 1926, mining companies became even more concerned about increasing the efficiency of milling operations. This concern for more efficient milling operations resulted in two significant developments. The first involved the re-milling of tailings, or waste products from previous milling operations. Beginning in 1925, the flotation process of milling ore began to see significant usage. Although not used as the only means to re-mill the tailings, the mine operators soon realized that the flotation process could be used to recover significant amounts of ore from the mill waste piles that had been growing during the years of inefficient milling operations. Milling efficiency had improved from approximately 60% in 1919 to almost 90% by 1937. Also, the minimum amount of lead and zinc required from a ton of tailings was less than required from a ton of mined crude ore because the cost of underground mining operations was eliminated.

Initially, small mills were built next to the tailings piles. As transportation facilities improved with the minefields, the chat piles began to be transported to central mills in the 1930s.[429]

In the Joplin district and the first fifteen years of the Picher field, almost every mine had a mill next to it. A relatively small amount of central milling occurred in the Joplin district before 1912. However, central milling was never prevalent because of the scattered location of the mines and inadequate facilities for transporting crude ore to a central mill location. The use of an independent mill for milling a mine operator's ore, called custom milling, was not successful because of mine operators' and landowners' fears of an inaccurate accounting of their ore yields and royalties.[430]

But both central milling and custom milling were ideas whose time had come by 1929 when the Commerce Mining and Royalty Company constructed the Bird Dog Concentrator. The mill's 100 tons per hour capacity produced a greater amount of concentrates per ton at a lower cost. Faced with marginal ores and stiff competition from mining in other parts of the country, mine operators and landowners were convinced, and the dominance of the individual mill was broken. The number of mills in the Miami-Picher dis-trict declined from 143 in 1926 to 83 in 1930.[431] Within twenty years after the Bird Dog mill's construction, custom and central mills performed 90% of the district's milling. The giant Eagle-Picher mill was built in 1932 and had an initial capacity of 3,600 tons per day that had grown to 15,000 tons per day by 1950.[432]

*Section of panoramic view of Evans-Wallower No. 8 Mill –
1935(Photo courtesy of Baxter Heritage Center and Museum, Baxter
Springs, Kansas)*

Central Mill tailings (chat) pile (Photo courtesy of Dobson Museum & Memorial Center, Miami, Oklahoma)

Blue Goose Mill (Photo courtesy of Baxter Heritage Center & Museum, Baxter Springs, Kansas)

Hand jig used in milling operation (Photo courtesy of Baxter Heritage Center & Museum, Baxter Springs, Kansas)

Section of panoramic view of St. Louis Mill (Photo courtesy of Baxter Heritage Center & Museum, Baxter Springs, Kansas)

The Eagle-Picher Mining and Smelting Company was the largest mining company in the Miami-Picher field followed in size by the Commerce Mining and Royalty Company whose Bird Dog mill was located two and one-half miles west of Picher's "A" and Connell Streets. Eagle-Picher's new mill was built on 700 acres halfway between Cardin and Commerce.[433] The construction of the Central Mill involved many pioneering design features to overcome complex and difficult operational problems. But there were other logistical and peripheral problems in addition to those associated with the actual construction of the mill itself. Some of these problems included purchase of rights-of-way for railroad spurs to haul ore from the various outlying mines; negotiation of agreements with Northeast Oklahoma and Frisco Railroad Companies for rental of locomotives to pull the ore cars over their tracks; purchase of specialized rail cars for loading, hauling, and dumping the ore; and purchase of trucks and special loaders for loading ore and moving it from smaller mine operations that dumped their ore on the ground instead of large hoppers. The Central Mill was constructed with five huge underground hoppers. Four were designed such that rail cars could be pulled over the hoppers by locomotives and dumped following weighing. The fifth hopper was built with a bridge for trucks to dump their ore loads picked up from small operators. Ore from only one producer was permitted in a hopper at one time to insure the mine operator and landowner would receive proper credit for the amount of ore sent for processing. Because of variations in the amount of concentrates in a batch of crude ore, a sampling procedure was developed to determine the percentage of lead and zinc contained in a hopper's batch of crude ore.[434]

The five hoppers had a depth of sixty feet and were built to allow the ore from only one producer to be dumped, crushed, sampled, and conveyed to surface hoppers. Below the five 500-ton hoppers were two forty-eight inch jaw crushers that would break all chunks of ore measuring as much as two feet in diameter to a size of no more than five inches in diameter. The jaw crushers were located thirty feet below the bottoms of the hoppers, or ninety feet below ground level. Once crushed, the ore would be transported the ninety feet back to the surface for sampling and dumping into surface storage hoppers. A sample of several tons was reduced to several pounds, crushed again into a pulp, and divided into three separate samples. One sample was sent to an assay laboratory chosen by the shipper, the second was analyzed by the Eagle-Picher assay laboratory, and the third was retained for sampling by an umpire in the event variations between the first two samples exceeded certain limitations.[435]

Before the lead and zinc could be separated into concentrates, additional crushing was conducted to reduce five-inch sized ore to no larger than one-quarter to five-eights inches in diameter. The crushed ore was then transported to a jig mill with a plunger in each cell to cause the heavier lead and zinc to gravitate to the bottom of the cell for removal. The rock, debris, and ore particles not removed would be sent to another jig mill that had water flowing through it as previously described. Following the jig mill process, remaining rock and ore were further crushed in ball mills that were large rotating eight-feet long cylinders containing five-inch diameter magnesium balls that would roll around inside the rotating cylinders crushing the ore and rock to a consistency of powder. From the ball mills, the powdery residue would be sent to the flotation plant.[436]

With Eagle-Picher's 1939 purchase of the third-largest mining operation in the Picher field, it was necessary for the Central Mill to enlarge its processing capacity as well as to continue lowering the per ton cost of milling. This was made possible by the introduction of a cone plant process. In this process ore was fed into a thirty-foot tall metal cone, large end up, and the addition of water and Ferro Silicon. Because of the differing densities of the rock and ore, the lead-free and zinc-free rock would rise to the surface and be removed to the chat piles. The cone process only worked if the rock and ore fed into the cone was between a five-eights and one and a quarter inch in diameter. Prior to the introduction of the cone process, all ore would have to be run through the milling process. This process eliminated about 80% of the rock, leaving only 20% to be milled, and resulted in a dramatic reduction in the cost of milling a ton of ore.[437]

The Tri-State Mining District has 300 miles of underground tunnels. Some say the mines are so interconnected that it would have been possible before the mines filled with water after the pumps were shut off to travel underground between Joplin, Missouri, and Picher, Oklahoma.[438] Others say the twenty-five mile distance was blocked by an ore-less limestone barrier between Baxter Springs and Galena, Kansas, which prevented passage between the Missouri and Oklahoma mines.[439] The vastness of some of the caverns left by the mining operations is illustrated by the size of the Eagle-Picher West Side Mine in Treece, Kansas, less than a mile from the Oklahoma border to its south. The 428-foot deep mine had a ceiling height of 125 feet.[440] The cavern's floor space was estimated at about five acres,[441] or over 200,000 square feet, more than enough to hold the 175,000 square feet of the United States Capitol Building's footprint.[442]

The visible legacy of mining in the Miami-Picher district is not the extensive maze of drifts and caverns beneath the surface but the enormous chat piles sprinkled over the forty-four square miles surrounding Picher. Mine tailings, or chat, is to ore what chaff is to wheat, and the men of the Miami-Picher district built mountains of chat. Following a 2002 visit to Picher and a climb to the top of one of the chat piles, then EPA Superfund program director Jo Ann Griffith stated, "All the pictures that came across my desk in Washington don't even begin to do it justice. I thought I would go to the top...and see the Tar Creek Superfund Site, but it just seems to stretch on forever, one chat pile after another. This is going to be the biggest job I've ever imagined."[443]

What Ms. Griffith may have not known is that the chat piles she saw represented only about 10% of the number of chat piles that once existed in the area. One can climb to the top of a 200-foot high chat pile just a couple of blocks east of Connell Street and "A" Street and count about twenty large chat piles remaining in and around Picher. Small chat piles and portions of formerly large piles not readily visible dot the remainder of the Miami-Picher district. Almost all of the 227 mills that existed at one time or other in the district had its own tailings pile. These mills served over 1,320 mineshafts sunk in the district.[444] A substantial portion of the enormous amount of mine tailings came from the ore loaded into cans one scoop at a time and hoisted to the surface one can at a time.

Although chat piles were considered waste material by mine operators, the flint rock chat produced by much of the Miami-Picher district was found to make a good base for roads as opposed to limestone chat that would erode and wash away. Railroads became the biggest buyer of the discarded chat and used enormous amounts. The chat provided economical ballast for rail beds after it was crushed, stockpiled, and waited for loading. The flint rock chat had an additional advantage in that it produced less dust when trains passed over it. Dust from the limestone chat, whipped up from the roadbed by a moving train, would be swept into the wheel bearings, causing excessive wear and a shortened life. The fine powdery tailings from flotation plants was not used for roads or rail beds but stored in float tailing ponds that can still be found in the Picher fields.[445]

The Central Mill had a chat loading facility capable of loading 106 eighty to one hundred ton rail cars per eight-hour shift. The cars were loaded two at a time from hoppers filled by trucks and conveyer belts. The Central Mill pile was called the largest chat pile in the world.[446] At one point in time the Central Mill accumulated fourteen million tons of chat due to a rail strike that prevented the chat from being hauled away.[447] At ten thousand tons loaded per a typical eight-hour shift, it would take between five and six years to haul away the mountain of chat without including the continuing additions to the chat pile during those years. One must also remember that chat was being hauled away from other piles in the Miami-Picher district at the same time.

No other chat pile in the district reached the height and size of the Central Mill's tailings. In aerial pictures taken in its hay day, the mill complex and long trains appear as small toys next to a giant sand pile. The Central Mill chat pile reached a height of approximately

400 feet, and its base covered eighty acres.[448] There were many other piles that reached heights of 200 and 300 feet.

One may begin to understand the enormity of the Central Mill chat pile when it is compared to the size of Egypt's Great Pyramid of Giza, estimated to have been built between 2500 and 2600 BC. The oldest of the seven wonders of the ancient world, Khufu took twenty years to build, reached a height of 481 feet (now 449 feet minus its cap stones), and covered thirteen acres. Combined with the two other Giza pyramids, the ground covered totals approximately twenty-seven acres compared to the Central Mill's eighty-acre pile. The Great Pyramid weighed almost six and one half million tons, less than one-half of the fourteen million tons of the chat stored at the Central Mill during the rail strike. The pyramid's peak exceeded the Central Mill chat pile's peak only by fifty to seventy-five feet. The combined weight of the three Pyramids of Giza is less than fifteen million tons.[449] The combined weight of the three pyramids is less than 10% of the 181 million tons of lead and zinc ore produced by the district's mines.[450] If the twenty-year estimate of the time it took to build the Great Pyramid is correct, then it would be reasonable that the other two adjacent pyramids would take an equal or somewhat lesser amount of time. Therefore, to build all three pyramids would have taken approximately sixty years or less to build compared to the fifty-four year life of the Picher field between 1915 and the end of mining operations in 1969. In reality, the chat piles were largely completed by the mid 1950s when many mines began shutting down. It is even more remarkable when one considers that not only was this volume of chat produced during those years, but also a great portion of it had been removed from the district.

Only in recent years has the myth been refuted that the pyramids were built by slave labor. The young men who built the pyramids were conscripts, healthy young men from small villages and farms scattered southward along the fertile Nile valley. The length of service, somewhat akin to a modern military draft, may have been as short a single wet season. Only a handful of skilled artisans, administrators, and supervisors would have remained with the project year after year. It may have taken a hundred thousand men to build the Great Pyramid, but it is also likely that only a small fraction of that number worked on the pyramid at any one time.[451] It is not possible to know or accurately estimate the number of miners that worked in the Miami-Picher field during the near half-century of its life. But given the number of mines and mills in the district, the number would have been in the tens of thousands, and like the pyramids, only a small portion of that number would be working in the Miami-Picher minefields at any one time. Over 11,000 men worked in the all three sections of the Tri-State Mining District in 1924, the peak year for employment. That number shrank to 1,800 in the Depression year of 1934. However, employment increased to 3,500 after 1935 and peaked again in 1953 at 4,100 before employment began its rapid decline as many mines began shutting down in the late 1950s and early 1960s.[452]

After considering these comparisons, one begins to realize the enormity of the accomplishment of the miners and mine operators of the Miami-Picher field. Although lacking the romance and esthetic appeal of the pyramids, the magnitude of the work accomplished by the men and a few women in the small corner of northeastern Oklahoma becomes evident.

Hell's Fringe

"Therefore I hated life; because the work that is wrought under the sun is grievous unto me: for all is vanity and vexation of spirit."[453]

Hell's Fringe was the name given by one writer to that area just inside Indian Territory border adjoining Kansas, Missouri, and Arkansas.[454] Before statehood, it was a safe haven for the outlaw, cattle rustler, murderer, and other misfits of society. These lawless pariahs' threats to life, limb, and property created a purgatory, a place of temporal punishment if not the edge of Hades itself, for the law-abiding citizen, white man and Indian alike. The outlaws of the territory in the late 1800s were noted in a previous chapter. Statehood brought a measure of law and order in 1907, but it did not end the lawless side of life, particularly in the wild mining camps and boomtowns of the Miami-Picher mining fields in far northeastern Oklahoma. It would be many years and a new generation or two before the frontier attitudes and mindset of both citizens and lawbreakers would change.

By the 1920s, outlaws no longer rode horses or sported revolvers hung from their hips. Rather, Hell's Fringe was populated by gangsters with machine guns and fast automobiles. Commerce, Oklahoma, tended to be a hangout for a number of infamous criminals of the era between the two world wars. Local citizens would occasionally see men and women whose names and pictures had made newspaper headlines throughout the country; names such as Pretty Boy Floyd; Clyde Barrow and Bonnie Parker; Machine Gun Kelly; Ma Barker and sons; George Devine; and Pascal (Pat) Ramsey and his attractive wife, Nadine. Bonnie Parker had lived in Commerce for a time. Several reasons have been advanced for Commerce's popularity with this criminal element. Some say Route 66, the nation's new highway stretching from Chicago to Los Angeles that passed through Commerce, made it convenient to other larger towns in the area.

TRI-STATE MINING DISTRICT—TOWNS.

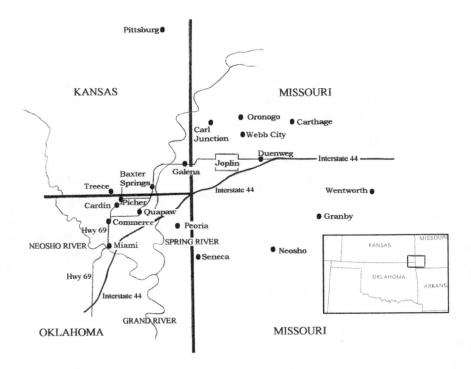

Figure 12

Others believe the heavily wooded forests to the east and large chat piles that surrounded the area attracted those with a desire for a safe haven from the law.[455]

On Easter afternoon, April 1, 1934, Bonnie Parker and Clyde Barrow along with Henry Methvin, a shy country boy recently released from a Texas Prison, murdered two Texas highway patrolmen investigating a parked car on the side of the road. Five days later, the trio became stuck on the Lost Trail road a half mile west of Commerce. The gang attempted to stop a passing motorist, but he fled when he saw the guns stacked inside the immobile Ford. Alerting Commerce Police Chief Percy Boyd and Constable Cal Campbell, the officers set out to investigate the motorist's story. As the officer's car approached, the Ford suddenly began backing up at full speed but ran into the ditch again. The offices stopped in front of the Ford, got out of the car, and began walking toward the parked Ford. Campbell noticed that Barrow, now standing next to the car, held an automatic rifle. Campbell drew his pistol and fired three shots. Barrow returned the fire, killing Campbell. Boyd fired four shots but fell to the ground when hit by a bullet. Chief Boyd was taken hostage but later released.[456] Methvin would be captured and, along with his father, assisted in setting up the ambush that killed Bonnie and Clyde on May 23, 1934. The State of Texas pardoned Methvin, but Oklahoma tried and convicted him for the murder of Campbell. Sentenced to death, his term was commuted to life. Methvin was granted a parole after serving eight years only to be killed when run over by a train in 1948.[457]

Even into the early 1940s, Commerce remained a magnet for the criminal element. Pat Ramsey had escaped from a Missouri prison and had joined up with his wife, Nadine, and another dangerous criminal by the name of George Devine. Aubert Sidwell, a longtime law enforcement officer and then Commerce police chief, received a call from Joplin police on May 16, 1942, that the Ramsey gang had just gone on a wild crime spree and were thought to be somewhere near Commerce. Sidwell called for reinforcements from the Miami Sheriff's Department and would soon be joined by three deputies. The four lawmen drove around Commerce and soon spotted the gang's car in front of Minson's grocery where the trio had gone to buy fruit. The gang jumped into their car and sped out of town with the lawmen in pursuit. At ninety miles per hour, the lawmen could stay close but couldn't stop them. Nadine tossed a quantity of roofing nails on to the road from a window of the get-away car in an attempt to puncture the officers' tires. At one hundred miles per hour it would take only eight minutes to cover the next twelve miles when the Ramsey car veered from the road and crashed into a tree. The three occupants escaped into a wooded area. A gun battle raged to a standoff, but Nadine Ramsey and George Devine eventually surrendered. Pat Ramsey was found dead among the trees.[458]

Not all encounters with the violent gangs of the era ended in tragedy. Mrs. Maude George was in her farm home near Afton, a town a few miles south of Miami. Alone after her husband had gone to the field, a car drove up with four armed men inside. The leader, a very presentable young man, assured her that she wouldn't be harmed. They asked Mrs. George to fix breakfast for them, as they were hungry. The men consumed the ham, eggs, red-eye gravy, homemade biscuits, and coffee. As they prepared to depart, the leader handed Mrs. George twenty-five silver dollars and said, "You can tell your friends you just served breakfast to Pretty Boy Floyd."[459]

The honest citizens of Ottawa County were not all as docile as Mrs. George. Around 1900, it was the custom that during haying season haying crews would go from farm to farm to put up the crop for the winter. A man by the name of Barker followed the haying circuit and brought his wife and young boys with him. While at the Frank and Melissa Booth farm, the Barker boys stole something from the Booth home. The Booth children, having seen the theft, promptly chased, caught the Barker boys, and retrieved their property. After finding out about the altercation, an angry Mrs. Barker gathered her boys and proceeded to the Booth house to punish the Booth children. Mrs. Booth was waiting at the yard gate with her Colt 45 and ordered the Barkers to never step foot in her yard again. Many years later Mrs. Barker would be known as Ma Barker of the Ma Barker gang that included her four outlaw sons and Alvin Karpis.[460] J. Edgar Hoover, longtime head of the FBI, would call Ma Barker, her sons, Karpis, and his gangster friends the smartest and most dangerous gang of hoodlums the FBI had ever dealt with. However, in later years, the role of Ma Barker as the mastermind of the gang would be revealed as a myth. She knew of the crimes, traveled with the gang, and lived off their ill-gotten gains, but as one of the gang members would later say, "The old woman couldn't plan breakfast." Apart from doing jigsaw puzzles and listening to *Amos 'n Andy*, watching out for the welfare of her sons was her main concern as it was when she intended to punish the Booth children. However, Hoover was right about the Barker-Karpis gang. Although basically a bunch of hillbillies, the scale and ambition of their crimes would far exceed those of the more notorious gangsters of the period.[461]

But the story of Melissa Booth and her Colt 45 did not end with her encounter with the Barker clan. Her husband passed away, and she rented the farm's pasture to a local sheriff for his cattle. But when he took his cows out in the fall, he did not pay the rent due Mrs. Booth. Again the following spring, the sheriff brought his cows to the Booth pasture for grazing. When he arrived in the fall to get his cows at the end of the second grazing season, Mrs. Booth met him at the gate and asked for two years' rent. The sheriff's laugh soon faded as Melissa Booth pulled the Colt 45 from beneath her apron and pointed it at him. The accompanying deputy was hastily dispatched to town to get the rent money so the sheriff could pay the gun-toting Widow Booth and retrieve his cattle.[462]

Crime imported by the big name gangsters of the era was not the only mayhem committed in the Miami-Picher mining district. In fact, most of the crime was of the local variety, and much of that involved alcohol. Liquor sales were prohibited in Indian Territory before statehood, and the ban was continued by a provision in the state's new constitution. But bootleg alcohol was plentiful due to local stills or was easily obtained in nearby "wet" states. And where better to make moonshine than in a deserted mine drift? A well known and successful mine operator by the name of Mike Evans built a huge still in one such abandoned drift. Evans sunk a shaft beneath a large garage-warehouse down to the abandoned drift on one of his leases. He installed an electric elevator in the shaft located behind a false wall at the back of the large building. Evans sold the mine lease in 1925 but reserved that part under his warehouse. It wasn't long before federal revenue agents raided the mine, discovered the still, and arrested the two new and unsuspecting owners. Mike Evans eventually confessed and was sentenced to six months in jail at Vinita, Oklahoma. Such was the era's casual attitude toward bootlegging that he was required only to sleep in the jail at night. Evans could leave during the day and go wherever he wished. Why did he do it? Mine operators were gamblers, always betting their next lease would be a bonanza. For Mike Evans, making moonshine was just an extension of the thrill and excitement he found in the mining gamble.[463]

For the average miner, the danger posed by the high-profile killers of the day paled in comparison with the dangers awaiting him everyday on the job. By far the greatest danger and the greatest cause of death in the mining fields were falling rock.[464]

The Grim Reaper chose a variety of other methods to help accomplish his work, and seldom a week went by without the newspapers reporting another death or injury in the minefields. A thirty-year-old hoisterman at the Eureka Mine was killed after falling 212 feet to the bottom of a shaft.[465] Two miners, riding in a can as they were being lowered to their jobs, were killed when an emergency brake on the hoister broke and dropped them 150 feet to the bottom of the shaft.[466] A twenty-three-year-old died of complications from a falling slab dislodged by a roof trimmer who had jumped to avoid the falling slab and at the same time threw his pry bar. The bar struck the victim in the side, and the slab of rock crushed his leg.[467] Nine miners were hurt and one killed in an explosion in the Lucky Jew Mine when natural gas accumulated in a natural cave was ignited by the lamps of the miners that had gone in to explore the cave.[468] Workmen were sinking a shaft at the Lawyer Mine when dynamite sent a large boulder into the air followed by a crash through the roof of a doghouse, a place where miners changed clothes before and after a shift. The falling boulder struck and seriously injured a hoisterman who happened to be in the doghouse at the time.[469] During a shift change at the Liza Jane Mine, a miner at the surface awaiting his turn to go down slipped and fell 297 feet to the bottom of the shaft. Four men had been lowered, and all but one had left the can when the falling man's body struck the man remaining in the can. Both men were killed.[470] Four miners fell to their death at the Aztec Mine in one of the worst accidents of the Picher field. The four men were being lowered in a can to the bottom of the shaft when the can struck the cribbing on the side of the shaft causing it to up-end.[471] One miner was killed and three injured at the Lawyers' Mine when a premature explosion occurred as a round of shot was being loaded in preparation for blasting.[472]

Most deaths occurred in the mines, but workers in and around the mills were not immune from death. A night watchman was burned to death in a fire that started in a compressor room breaker box. It was reported that the watchman probably received a severe shock when he attempted to shut off the current. Flames flashed from the breaker box and set the building ablaze and the watchman was unable to escape.[473] A general ground foreman at the Wade Zinc Company was killed when an emery wheel being operated at high speed broke. A piece of the wheel struck the man's head, fracturing his skull, destroying an eye, and lacerating his face.[474] Another mill worker was killed when his leg became entangled in a cable and shive wheel.[475]

Not all death in the mines involved miners. The body of a Neosho, Missouri, man who was visiting his uncle was found at the bottom of the St. Louis No. 11 mine. Both men had been drinking, and authorities speculated that the dead man had either crawled under the railing around the shaft or had been thrown in by his uncle following a possible argument.[476]

The story is told of Charles B. Osborne, better known as Uncle Charley, who was reported to be the last man killed in an underground accident in the Miami-Picher field prior to the shutdown of the last mine in 1969. Uncle Charley began his pick and shovel work in the mines as a teenager near the beginning of mining in Ottawa County. Rising to the level of mining superintendent, he was forced to retire in the mid 1950s and moved to Texas. By 1967 and in spite of family opposition, the seventy-six-year-old Uncle Charley had returned from Texas and obtained a lease on the Brewster Mine at Zincville. He argued that he had spent all of his life underground and couldn't think of a better place to die. While down in the mine with his brother to survey its condition, he remarked at how good and happy he felt to be back in the mines. He then took off his hard hat to wipe the sweat from his face. At that moment a small rock no larger that a half dollar fell from the roof and struck Uncle Charley on the head. He died instantly, his last wish being fulfilled.[477]

For every death in the mines, there would be many close calls. One such occurrence happened when five miners decided to do their end of shift blasting fifteen minutes early as it was one of the men's wedding night. After lighting the fuse, the men rushed to the mineshaft to be pulled to safety by the hoisterman. However, the hoisterman had left his post to retrieve a sandwich from his dinner bucket, thinking he had fifteen minutes before the men would need to be pulled from the mine. When the men discovered the hoisterman's absence, they began climbing the cable, hand over hand, to a niche in the shaft wall below where it had been cribbed. Four men reached safety as the fifth climbed the cable. Halfway up, the miner who was to be married that night lost his grip and fell back to the shaft floor. One of the men climbed back down the cable and rescued the fallen man, arriving at the niche just as the first explosion happened. Near suffocation from the powder smoke, the men were eventually pulled from the shaft, and the groom-to-be spent several months in the hospital. However, at the insistence of his fiancée, he and the young woman were married in his hospital room five days later with his buddies in attendance.[478]

Another close call occurred as fifty-four miners were pulled from the mines at the end of the shift. The hoisterman had sent the can down fifteen to twenty times to pull the miners up at two to four per trip. The fifty-fourth man stepped from the can, and the can was raised the final few feet to be hung in the screen room to await the next day's work. At that moment the bail holding the can to the cable hook broke, and the empty can that had hoisted tons of ore and fifty-four miners to the surface that day hurtled to the bottom of the shaft.[479]

People who have a heightened fear of the unknown and/or a greater than normal chance of death tend to develop defenses or practices that allow them to cope with their circumstances. These defenses and practices may include a reliance on one's faith or perhaps the development of a fatalistic attitude that future events are fixed and humans are powerless to change them. Others develop superstitions, a strong belief in the cause and effect relationship between unrelated actions and events.

Miners certainly had reasons to fear the unknown and risked a much greater chance of death than most. Tri-State miners developed a series of superstitions to help them cope with their perilous conditions. Few of these had any basis in fact. If a man was cross-eyed, he would not be allowed in the ground because he would bring bad luck. A careless miner would have an accident unless his father was a preacher. A buzzard flying over a mill was a harbinger of a fatal accident. A penny thrown into a shaft before going down would bring good luck (but not for the miner at the bottom of the shaft struck by the tiny copper missile). A mine named after a woman would not pay off.[480] This superstition may not have been taken too seriously, given the fact that many successful mines in the district were named after women.

Women in the mining towns and surrounding countryside did not face the daily dangers of their husbands and sons. But in many respects their lives were just as difficult and perhaps more so. If we could peek behind the flower print cotton feed sack curtains of a typical miner's home in the Picher field, we might gain a better understanding of the hardships and often hopeless struggles a miner's wife encountered on a daily basis. The small frame house she occupied was rented, usually needed paint, and couldn't be bought even if they could have afforded it as it was on restricted Indian land. The dwelling of three, four, or if lucky, five rooms would house her, her miner husband, and their children. Water came from a barrel and was hauled in. The outhouse would lie beyond the vegetable garden planted to supplement a diet of beans and potatoes. Dirt, dust, scorching heat, and bitter cold were her enemies. She constantly feared hearing the long wail of a mine whistle announcing an accident at the mines. When she heard it, as she often did, she would breathe a prayer, "Lord, let it not be my husband."

—

Where would she live? How would she feed her children? Perhaps she could get a job cooking at one of the boarding houses. Soon she would be silently embarrassed by thoughts of her own survival if her husband were maimed or killed. She regretted the harsh words she yelled as he left that morning, berating him because of the loss of a week's wages during last Saturday night's drinking binge. Would her thirteen-year-old son follow his dad to the mines in two or three years? She couldn't bear the thought, but that new boy who passed her porch this morning, he couldn't have been too many years older than her son. He said he was from Kansas and that it was his first day in the mines. She saw her worried face as it stared back from the mirror. She was only thirty-two, but looked fifty. Would her daughters marry miners and look old before their time too? Then she heard it, the familiar step on the front porch. Thank God, it wasn't to be her man killed this time. But his worried face revealed that it was someone she knew. Her unspoken question was answered by the piercing scream followed by the anguished sobs of her neighbor. That night as she lay next to her exhausted husband listening to his dust-filled lungs as they wheezed with effort, she again thanked God for sparing his life one more day.

—

With a few exceptions, mining was a man's world, but women would make his hard life bearable. But some women assumed roles other than the typical life of a miner's wife. Mr. and Mrs. O. J. Horn came from Tennessee in 1917 with a small team of mules and contracted to haul ore from the Mahutoka Mine. Mr. and Mrs. Horn each drove a team to pull the ore cars from the mineshaft to the mill. In 1918, Mr. Horn became a hoisterman and taught his wife to hoist cans from the shaft. With her husband laid up for two weeks with an illness, Mrs. Horn did his job as well as hers. Other early mining camps saw women handling men's jobs. Two women worked as coal heavers shoveling coal from rail cars into a chute. Coal was used to fuel the steam engines that ran the mining machinery. The 105-pound and 129-pound women earned fifteen cents a ton. Clara Platter worked in the Peoria mines as a girl of thirteen.[481]

Emma Semple would have been considered a remarkable woman

even in today's modern world. But her drive and success during the early years of the twentieth century when fewer opportunities existed for women only adds to one's admiration of this woman and her accomplishments. She moved from Galveston with her parents after floodwaters destroyed the family farm. Settling near Chicka-sha, Oklahoma, she became a champion cotton picker while in high school. She later married but was left a widow with a four-year-old son. To provide for herself and son, she attended an Oklahoma City business college and graduated in 1914. By 1918 she had moved to the Miami-Picher area, becoming the office manager and co-owner of the Early Bird Mine. Emma and partner C. Y. Semple operated the mine for twenty-nine years, and it would become one of the richest in the Tri-State district. Such was her concern for the well being of her employees that it was Emma's practice to dress in boots, pants, and hardhat and go down into the mine each morning for a safety inspection before the miners arrived. Semple's wife died in the early 1930s, and he and Emma were married in 1938. She was widely known for her kindness and generosity to needy families and often paid for school and college for those that needed it. After the death of Emma's son and his wife, Emma adopted and raised her grandchildren.[482]

Other children were not so fortunate. Chickens scratching in a straw pile near a mule barn in a mining area just south of Picher in early December 1917 uncovered an abandoned newborn baby. Although badly burned by the heat generated from the manure pile, the warmth probably saved the baby's life from exposure to the season's cold tem-peratures. The emotions of the women of the area were torn between a wish to rain retribution on the head of the unknown mother and their desire to care for the unfortunate baby.[483]

Mining accidents occurred often, suddenly, and usually resulted in severe injury or death. The miner would never know when he would descend to the mines or walk into the mill for the last time. But given the amount of horseplay among the mining men, an uninformed observer would never have guessed the dangers that awaited them. Many pranks were played on men new to mining, such as advising a new man to wear a corset when shoveling ore to keep from breaking his back. Other new miners would have the lids to their lunch buckets soldered shut while they were working.[484]

Much of the fun was had during their underground dinner break (dinner was at noon as supper was the evening meal). Often the miners' humor would involve a "kangaroo" court. In addition to the judge, there would be a prosecuting attorney, assistant prosecuting attorney, secretary-treasurer to collect the fines, and a court stenographer to administer oaths on a copy of *Whiz Bang*. Defense attorneys were not allowed. Offences might include laying one's hat on the bench, impersonating a miner, gambling, attempting to induce one miner to bend a steel bar around another miner's neck, being a snitch, spitting into a downdraft, or any other trumped up charge the men could think of to separate a miner from a fine of up to twenty-five cents. Court fines were given to injured or needy miners and their families.[485] Some miners took their courts a little more seriously. The Lucky Bill Mine had a list of twenty-five infractions that would cost offending miners five cents and a couple of licks with a paddle. Many of the rules had a sound underlying reason behind them. Examples include riding up or down the shaft when smoking or with lighted lamps while shots were in the can, smoking while handling explosives, various sanitary offenses, and scuffling in the can or around the shaft. The small fines amounted to over two thousand dollars at the Lucky Bill mine during the two-year period between 1923 and 1925.[486]

It was the custom in the first half of the twentieth century for corporations to sponsor employee amateur sports teams. Phillips Petroleum Company was well known for its 66ers basketball team. Phillips' sponsorship of the team began in 1920 and lasted until competition from professional teams led to the 66ers being disbanded in 1968. The teams had a remarkable record of 1,543 wins, only 271 losses, and included eleven Amateur Athletic Union and two Olympic Trial championships. Thirty-nine of the team's players were AAU All-Americans, and four company presidents and many other top executives played on the team over the forty-eight year life of the 66ers.[487]

The mine operators of the Miami-Picher field also sponsored company sports teams, and the sport of choice was baseball. Frank

Phillips knew little about basketball, but he saw the potential advertising and goodwill value, as well as a means of developing executive skills in the young men.[488] While the Phillips' teams involved only a handful of company employees that would become future executives in the company, the baseball teams of the mining district involved hundreds of employees and were established to build employee morale. Eagle-Picher sponsored a league of twenty miners' teams, each from one of the company's mines in the Picher field. The company built a baseball field and grandstand at the entrance of its Central Mill between Cardin and Commerce.[489]

Eagle Picher miners—Mutt Mantle (Mickey Mantle's father) is standing in the back row in front of the wooden column between the two buildings (Photo courtesy of Baxter Heritage Center & Museum, Baxter Springs, Kansas)

Mickey Mantle—Miner (Photo courtesy of Baxter Heritage Center & Museum, Baxter Springs, Kansas)

Barney Barnett was the ground boss at the company's Barr Mine and was known for hiring "baseball miners," employed more for their prowess on the diamond than their skills and work ethic below ground. So good were his teams that company executives would make periodic unannounced inspections of the Barr Mine and fire the "baseball miners" found sleeping or hiding behind pillars so that other teams would then have a chance of beating a Barr Mine team composed of real miners.[490]

As we have seen, the Picher mining field was a mixture of the famous, infamous, but mostly the unknown. One of the few in the first category was Mickey Mantle, the legendary baseball player with the New York Yankees of the 1950s and 60s. Mickey grew up in

Commerce, Oklahoma, and married a Picher girl whose father managed a lumberyard in Picher. Mutt Mantle was Mickey's dad and the ground boss of Eagle-Picher's Blue Goose No. 1 mine. In addition to coaching his Barr Mine teams, Barney Barnett coached a group of fifteen and sixteen-year-olds, including Mickey Mantle, that beat most of the area teams in the Tri-State.[491] Mickey's grandfather also worked in the mines, and as young men, both Mickey's father and grandfather had pitched in the semi-pro leagues on weekends. The two began teaching Mickey to switch hit when he was only four years old. By the age of fifteen, Mickey had begun working in the mines with his father and playing baseball with grown-ups.[492]

Mickey was playing for the D ball Joplin team when his play caught the eye of a scout and was called up to the Yankees for the last two weeks of the 1950 season. He would not have any playing time before returning to Commerce for the winter. That winter he would work in the mines for the last time.[493] He drilled holes and installed motors to blow fresh air into the mines. He rode the can down 300 or more feet to fix, clean, or remove the motors.[494] The next year would be one of the most eventful of the baseball player's life. During the 1951 season he was sent back to the minor leagues to regain his confidence and batting stroke. Mickey was feeling sorry for himself and not sure he would make it in baseball. Mutt picked up Merlyn Johnson, Mickey's sweetheart from Picher, and made the five-hour drive from Commerce to Kansas City where Mickey was. Mutt didn't extend the sympathy Mickey wanted but began packing Mickey's suitcase and telling him he could go back to the mines with him. It was the right message, for Mickey changed his attitude and was soon back with the Yankees. That fall he broke his kneecap in the second game of the World Series. Mutt Mantle was helping his son out of the taxi at the hospital when Mutt collapsed. Both shared a hospital room. Mutt was diagnosed with Hodgkin's disease and sent home to die. Mickey and Merlyn were married December 23, 1951, at her parents' home a little over a month after the cast was removed from his leg. Mickey and Merlyn began married life by living in Dan's Motor Court on Commerce's Main Street. With his earnings from the World Series, Mickey made a down payment on a home for his parents, their first with indoor plumbing.[495]

On May 6, 1952, Mickey's mother sent word to Yankee Stadium that Mutt was dead.[496] Mutt Mantle was only thirty-nine years old. Mickey's grandfather and two uncles had also died of Hodgkin's disease, none living past their early forties.[497] Mickey also lost a son to the disease. Diagnosed at nineteen, Billy Mantle lived seventeen more years and become addicted to painkillers and alcohol. Mickey Mantle had escaped the harsh life of a miner but descended into another pit of self-destruction including two divorces, fighting, gunplay, and alcoholism.[498]

Mickey was sixty-three when he died, almost a quarter of a century older than the age at which his father had died. A few weeks before his death in August 1995, Mickey called his former teammate and good friend Bobby Richardson to tell him that he had committed his life to Jesus Christ. Richardson had shared his faith with Mickey many times while they were Yankee teammates. Because of Mickey's decision, Richardson said that he knew he would see his teammate again.[499] Mickey Mantle had made it from Hell's Fringe to Heaven.

Miners' Health and Labor Strife

"Then I looked on all the works that my hands had wrought, and on the labour that I had laboured to do: and behold, all was vanity and vexation of spirit, and there was no profit under the sun."[500]

Concern for miner health and safety was virtually non-existent in the early years of mining in the Tri-State. It was only after 1900 that the nine-hour workday, six-day workweek would end. In 1899, Missouri lawmakers passed legislation to restrict miners to an eight-hour workday. The Oklahoma Constitution of 1907 limited work in mines to eight hours per day, but Kansas would not pass legislation to limit the miners' workday to eight hours until 1917.[501] It would be 1919 before the Oklahoma legislature would pass a bill commissioning the first study of workers' compensation.[502] Even a safety measure as obvious as equipping miners with hard hats would not occur until the 1920s and 30s. The first miners wore soft hats that protected their heads from dripping water and provided a place on the front of the hat to attach a carbide lamp. The first hard hats were made of "bakelite" and would be worn directly on the head. Later plastic and metal hard hats would provide an interior webbing to suspend the hat away from the scull and afford better protection and a more comfortable fit.[503] Although the hard hat would be of no use to those miners crushed by falling slabs of rock and dirt, it offered protection from severe injury or loss of life from the far more prevalent number of smaller falling rocks, such as the half-dollar size stone that killed Uncle Charley. It is interesting to visit the various museums of the early mining fields and observe the group pictures of miners. From these pictures, it is easy to pinpoint those years of transition from the miners' soft hat to the hard hat. Not only did the hats change, the early foul-smelling paraffin-filled "sunshine" lamps with lighted wicks were replaced by brighter and safer electric torches. Another constant in those pictures of miners posing in front of the mines were the crude drawings of Safety Pete, an early form of the safety poster that are now found in almost every office, shop, and factory in America.

Through passage of workers' compensation laws and a desire to lower insurance premiums, mine operators would eventually see the wisdom of providing a safer work environment. Facilities in which a miner could shower and change into dry clothes; electrically lighted mines; safety goggles; safety shoes; greater attention to dust prevention; and provision of health clinics, doctors, nurses, and regular physical examinations would contribute significantly to the health and welfare of the average miner. Unfortunately, most of the changes would not occur until after the boom years beginning in 1915 lasting until the late 1920s.

During the first three decades of the last century, a miner in the Tri-State had two choices with regard to his and his family's health and welfare: either he worked or he and his family would literally starve to death. And for a miner the only work to be had was in the mines. But to spend one's life in the mines and breathe the dust-filled air was to accept another kind of death sentence. Those who spent many years in the mines and were fortunate enough to escape the falling slabs and overturned ore buckets would very likely develop a case of "miners' con," as it was commonly called. Miners' consumption, or silicosis, is a dust-induced respiratory infection identified in European miners as early as the fifteenth and sixteenth centuries. But early miners of the Joplin district blamed miners' con on damp drafts, lack of fresh air, and smoking hand-rolled cigarettes.[504] Although ignorant in the area of science and health, miners intuitively knew those conditions affected their health. All of those supposed reasons for their ill health contained elements of truth and would certainly aggravate the health of a miner afflicted by silicosis. Miners were constantly pumping water from the mines as well as wetting down rock and ore to be loaded in the cans. Miners would wade through several inches of water on the drift floor and be soaked by water dripping from mine ceilings. Walls, ceilings, and floors were constantly damp. The cribbing surrounding the shaft walls would be slimy and slick from the constant presence of water. It was also easy for the miner to blame the dust-choked drifts as lacking fresh air when he really meant dust-free air. And certainly the smoking of tobacco, hand rolled or otherwise, would only tend to further weaken lungs damaged by silica dust.

Hard rock miners of the Tri-State would develop silicosis as they breathed dust containing silica, a constituent of sand, rock, or quartz. Silicosis occurs when dust particles as small as one five thousandth of an inch reaches the smallest air passages and sacs in a lung. As the dust accumulates, the lung tissues become fibrous, thicken, and scar. As the lungs become less elastic, breathing is impaired resulting in less oxygen being supplied to the blood. Decreased working ability, coughing, weight loss, weakness, and respiratory infections follow. High levels of dust exposure can result in an acute case of silicosis in as little as ten months. Silicosis cannot be reversed, and only the complications arising from the disease may be treated.[505] A miner would often attempt to hide his condition for fear of losing his job, but the persistent, hacking cough and diminished working ability would be obvious to his family and fellow miners.

The mine operator could easily replace rather than treat a lone miner afflicted with silicosis. But it would be tuberculosis of epidemic proportions in the Joplin area of the district that would cause mine operators to establish a Safety and Sanitary League on November 28, 1914, for education about and treatment of the disease.[506] Silicosis was not contagious, but tuberculosis could easily and rapidly spread through an entire work force and their families. In 1915, the Missouri legislature passed a mining law mandating that all mine operators install sanitary toilets, drinking fountains, facilities for workers to change clothes, and adequate water to abate dust created by drilling and blasting. Similar laws were enacted in Kansas and Oklahoma. Conditions for miners improved significantly over the next three years, but the incidence of silicosis and tuberculosis did not fall from previous levels. In 1924, the Bureau of Mines again pressed operators to provide an annual physical examination for all miners. That same year the Tri-State Zinc-Lead Ore Producers' Association, United States Bureau of Mines, and the American Legion's

Picher Post established a clinic to provide examinations. The clinic was expanded in 1927 to include an X-ray room, venereal disease clinic, and dispensary.[507]

In the twelve months between July 1, 1927, and June 30, 1928, over 7,700 men employed in and around the Picher mines were given physical examinations, including a chest X-ray and laboratory work. In the United States, an arbitrary three-stage classification of severity was used to classify a miner's silicosis. The first stage indicated definite physical signs of damage to lungs from exposure to dust. A reduced capacity for work resulted in a stage two classification. Stage three involved a serious or permanently impaired capacity for work. Over 20% of the mineworkers tested were found to have some stage of silicosis. Although others may have had some effects of dust exposure, the symptoms were not sufficient to be classified as having silicosis.[508] Those miners evidencing first-stage silicosis were warned, and those with second-stage silicosis were not recommended for re-employment.[509] The examination programs and efforts to get infected miners out of the mines before the silicosis became worse reduced both the frequency and severity of silicosis among the district's miners.

For many years and in spite of all efforts, Ottawa County would have the highest tuberculosis mortality rate in the United States. A 1936 school tuberculosis testing program in the Cherokee County, Kansas, mining towns of Baxter Springs, Treece, Columbus, Galena, and Riverton revealed that over 36% of 1,080 students and teachers tested positive for tuberculosis with almost 54% of those over twenty years old testing positive.[510] Seven years later, a test of almost 1,500 Picher schoolchildren resulted in a 20% positive test rate.[511]

For twenty-three years, Ruth Hulsman personified the efforts to improve the health and welfare of the miners and their families in the Miami-Picher mining field. Hulsman arrived at Picher in 1926 and was the area's first public health nurse. She remained until 1949 when the county took over her work on community health and sanitation. The Tri-State Zinc and Lead Ore Producers' Association was her employer and provided a budget to carry on her work that became the focus of every charitable work in the district. When she arrived in 1926, typhoid, smallpox, and diphtheria raged, often closing schools. In addition to clinic duties, home visits, and public school visits, public nurses such as Hulsman arranged hospitalization and treatment for the area's children. She drove hundreds of children with tuberculosis from Picher to the Talihina Sanitarium in southern Oklahoma. Making the long arduous journey, Hulsman often used her own car to drive crippled children with harelip, clubfeet, and numerous other deformities to Tulsa and Oklahoma City hospitals for surgery and other treatment or to schools for special training.[512] One of those children she helped was nine-year-old Mickey Mantle who suffered from osteomyelitis, an infection of the bone and bone marrow.[513] Perhaps without Ruth Hulsman's love, care, and concern for the children of the Picher area, the baseball legend that Mickey Mantle became may never have existed.

The Depression years added to the burdens of the public health nurses. Again, the Tri-State Ore Producers' Association funded Hulsman's efforts. Over 500 pairs of shoes were purchased and distributed to children in the district, soup was provided to schools, and a myriad of other activities were instituted to alleviate some of the poverty and suffering of the district's children.[514] In January and February 1935, a report derived from information supplied by the Oklahoma State Department of Public Works indicated that Ottawa County had 4,457 relief cases. This same report indicated that each case represented four persons on relief. In other words, the report indicated that almost 18,000 people were on relief in Ottawa County, slightly less than 50% of the county's 38,543 population recorded in the 1930 census.[515]

The depravations of the Depression years coupled with the growth of the American labor movement would briefly focus the nation's attention on the harshness of the living and working conditions in the Tri-State mining camps during the 1930s. However, the labor organizers in the Tri-State mining region generally did not find fertile ground in which to plant their organizations. Several reasons have been advanced for the reluctance of the Tri-State lead and zinc miner to embrace unions: a poor-man's camp mentality, the native-born character of the miners, and the absence of a diverse work force.

The remainder of the eighteenth century following the Civil War was an era of instant millionaires, every man his own capitalist. These boisterous, self-reliant, risk-taking adventurers crossed the Mississippi River and spread across the plains all the way to the California gold fields. Gold, silver, oil, lead, and zinc were the fabulous prizes awaiting the bold, quick, and strong. Many cared not for what lay under the land but were desirous of the land itself. Free land was available. All one had to do was claim it, bust the sod, plant the crops, pray for rain, stay on the homestead for the allotted time, and survive desperadoes, drought, floods, and ravaging hoards of insects. And, if the patch of land happened to be in the path or near one of the railroads rapidly spreading across the west, these would-be farmers would become business owners or land speculators. This Wild West mentality was well established on the lawless frontiers of the three states converging at the northeast corner of the Quapaw reservation. This spirit of self-reliance and chance taking was nowhere more evident than in the small mining camps of the Tri-State.

The two-man mine operators dominated the district prior to 1910. They conducted all of the elements of mining: performed elementary geology, prospected, built and operated a churn drill, sampled cuttings, operated air drills, used dynamite, operated water pumps, cribbed shafts, shored up mine drifts, shoveled ore, milled crude ore, repaired equipment, and built the mill and hoist structures.[516] These entrepreneurs had no need to join themselves together to gain a fair deal from the operators for they were their own bosses. But as the ore near the surface was depleted, the mines became larger, deeper, more mechanized, and considerably more expensive to finance, develop, and operate. More men were required to operate the mines. As a result, the average miner-operator evolved into an employee of a larger organization and became a specialist such as a hoister-man or a shoveler. Whereas the most valuable assets of the miner-operator in the poor man's camp were a strong back and a good work ethic, the larger and more complex mines required large amounts of capital, mining engineers, specialized and complex equipment, and a pool of willing and able workers. This division of labor and management came to the Tri-State by the end of the first decade of the twentieth century. The discovery of massive amounts of lead and zinc in the Miami-Picher field quickly sealed the fate of the small mine operators. But while the division between management and labor grew, the miners' resistance to efforts to organize for the purpose of collective bargaining continued. The gambler's hope still rested in the bosom of many miners that one day they could make that leap from being just an employee to becoming an operator, an employer of others.[517] This notion would persist with many miners until their failing health and the grim realities of the Depression shattered those long held dreams.

The homogeneous composition of the area's native-born white population added to the anti-union mindset as blacks and foreigners were often associated with unionization efforts. This perception was fostered by the strong presence of the Ku Klux Klan in Oklahoma and Kansas of the 1920s and 1930s.[518] The small number of foreign-born and blacks is evident upon review of the 1930 U.S. Census of Population for the four counties comprising the Tri-State mining district. Of the 170,768 inhabitants in the four counties, only 1,983 were black and 2,479 were foreign. The census data showed that Ottawa County, Oklahoma, containing the Miami-Picher lead and zinc field, had only 252 foreign-born whites and two blacks out of a population of 38,542. The absence of foreign-born and blacks in Ottawa County is even more striking when its population of those groups (254 out of 38,542, or 0.7%) is compared to the foreign-born and black residents of the other three mining district counties in Kansas and Missouri (4,208 out of 132,226, or 3.2%).[519]

Hand-in-hand with the lack of foreign-born minorities, another obstacle to unionizing the Tri-State miners was the native-born character of the miners. The Ozark hills of Missouri and Arkansas were a vital source of men to fill the mining camp rosters. According to one mine operator's view of the Ozarkian nature of the Tri-State miner, "All you have to do to get fresh miners (from the Ozarks) is to go out in the woods and blow a cow horn."[520] Others would dispute the stereotypical hillbilly origins of the Tri-State workforce. But the rural, agrarian independence of the Tri-State miner stood in stark contrast to the interdependence of union members elsewhere. Instead of joining unions to protect and provide for themselves during times of low wages or no wages because of layoffs or shutdowns, the miners would return to their roots and supplement their income with farming activities on the side.

Labor organization and collective bargaining activity in the Tri-State mining district prior to 1900 were sparse at best. In March 1872, a group of small, independent miner-operators gathered, elected a president, and drafted a code of conduct in dealings between the miners, landowners, and purchasers of the ore. The Miners' Benefit Association was formed in March 1874 to provide relief for miners and their families experiencing hardships. A strike in July 1874 against the Hannibal Lead Works in Joplin resulted in three days of violence and property destruction by miners protesting restrictions on ore sales in the open market and excessive royalty payments. The Western Federation of Miners and the I.W.W. began organization efforts after 1900. The Western Federation of Miners' 600 members went on strike in 1910 to gain better working conditions and higher pay. The membership had grown to 3,000 members by 1914, and the members struck again in June of 1915 for improved working conditions and pay raises. However, the operators broke the strike within three weeks.[521] Organization efforts in the Oklahoma portion of the district began between 1910 and 1914 with the smelters receiving ore from the mines. Most of the local unions survived less than two years. The Western Federation of Miners would become the International Union of Mine, Mill, and Smelter Workers.[522] In April 1925, the International Union was able to gather 3,000 miners for a meeting in Picher, but the union was not successful in establishing an organization.[523]

Tri-State miners had a national reputation as being difficult to organize as well as providing a pool of strikebreakers that could be recruited and shipped to other parts of the United States. Such was the district's reputation that Joplin was called the "scab incubator" in the November 1901 issue of the *Miners' Magazine*.[524] However, two events would occur that would give labor leaders new hopes of organizing the miners of the Tri-State lead and zinc fields: The Great Depression and the passage of pro-labor legislation by Congress. The Depression had been hammering the American worker and businessman alike for over three years when Franklin Roosevelt took office in early 1933. Congress quickly enacted landmark legislation to elevate the status of the American workers and enhance their ability to bargain collectively.

At the Tri-State lead and zinc district's peak in 1926, employment of miners and mill men totaled 10,200 with ore concentrate production revenues totaling $55 million. By 1932, employment had declined to 2,100, and ore concentrate production revenues had dropped to $4 million. Shovelers earned only $4.00 to $4.50 per day in 1935 while machine men earned $3.75 to $4.75 per day. Common laborers earned from $2.50 to $2.80 per day in 1935. The price per ton of zinc dropped from $55 in 1925 to $14-$18 in 1932.[525] During the depths of the Depression in 1932–1934, a sampling of the wages of miners and mill men working at the Eagle Picher mines and miles reflected an average weekly wage of $10.31, or $2.06 per day.[526] To the thousands of Tri-State miners who had lost jobs or watched their wages decline during the Depression, their reason for resisting union membership seemed far less important than it had in the boom years of the 1920s.

Pickets, Pick Handles, and Police

"For they eat the bread of wickedness, and drink the wine of violence."[27]

By 1933, eighteen years had passed since the first major discovery of lead and zinc in Picher. Deplorable housing, chronic health problems from breathing mine dust, loss of jobs, the insecurity of those with jobs, and the realization that only a few would reap anything other than day wages were the worries that had replaced the miner and his family's dreams of a better life. During this time of hopelessness and despair, a young man arrived in the Tri-State in the fall of 1933 and began preaching that a better life was possible.

Roy A. Brady was sent by the International Union of Mine Mill and Smelter Workers to once again attempt to organize the Tri-State lead and zinc miners. Brady's message was delivered with the zeal and fervor of an itinerate evangelist. He promised that union membership would result in better wages, more work hours, and improved working and living conditions. Brady said that those miners not joining the union would be prohibited from working in the mines when the International Union established the closed shop. He also said that workers were required to join the union under regulations of the National Industrial Recovery Act.[1] Brady was a brilliant orator and successful in attracting thousands to union membership. But Brady's success was a house built of cards, not the brick and mortar necessary to fulfill the promises made. Many not involved in mining were allowed to join the union. He lied to the miners regarding their obligation to join the union, as noted above. Brady blamed all the ills of the miners on the mine owners and operators without regard to other economic forces affecting the national and world economies. In spite of all his rhetoric, Brady never made a demand on the employers for better wages and working conditions. Brady's theft of the union's organizational funds would be his final insult to the trusting, desperate miners and their families, dealing a great blow to those fledgling efforts to organize the district's mine and mill workers. Following Brady's departure, International Union president Thomas Brown rushed to the Tri-State to rescue the struggling union locals.[528]

Employment in the district's lead and zinc mines had grown from 2,100 in 1932 to 4,800 in 1935, and the price of ore concentrate had risen to $30 a ton. By May 1935, Tri-State lead and zinc union membership was estimated to be four thousand[529] to five thousand.[530] Many in the ranks of the union were unemployed; however, a significant number of the 4,800 employed miners and mill men were unionized.[531]

The union's strategy had been to seek union recognition from individual operators of the mines and mills where union members represented a significant portion of the workforce. The mine operators initially tolerated the unions because of their belief that they would be able to maintain an open shop. But as the union began encouraging its members to initiate lawsuits for work-related illnesses and injuries, some operators either fired or threatened dismissal if a worker joined the union. Operators ignored the union's charges of discrimination and continued their refusal to negotiate with union representatives.[532]

In response to the International Union's failure to bring individual operators to the negotiating table, leaders of the six local unions met in March 1935 and drafted a proposal that two negotiating committees be formed, one representing the local unions and the other representing the operators. The two committees would be directed to reach an agreement by collective bargaining. The handwritten letter was ignored by the operators, as had the previous attempts to engage them in negotiations.[533]

The general mindset of the operators was probably best expressed by M. D. Harbaugh, the secretary of the Tri-State Zinc and Lead Ore Producers' Association. In a presentation to the Annual Metal Mining Convention, Western Division of the American Mining Congress on September 25, 1935, Harbaugh depicts the operators as viewing the bond between Tri-State miners and operators as close because of the common background shared by both. Harbaugh cited "the democratic attitude that has always existed between management and workmen." He stated that the operators and management truly believed "that the workmen saw no need for a union to do for them what they could do directly themselves." Harbaugh believed the strike had resulted as "an offspring of disordered minds dragging credulous men into a maelstrom of trouble out of which they were somehow led to believe they would be delivered into the promised land…"[534]

When Thomas Brown arrived in the Tri-State in April 1935 to deal with the mess left following Brady's theft of union funds, he inherited a union membership that had been promised much and received nothing in return for their dollars and trust. The six local unions' inept appeal to the operators had just failed. Without some action, Brown knew that the union would lose its members and cease to exist. Under Brown's leadership, the unions approached the Tri-State Ore Producers' Association whose members were the mine, mill, and smelter companies with whom the union had previously attempted to engage in direct negotiations. As would be expected, the Ore Producers' Association declined. Finally, the unions appealed for help from the United States Department of Labor's Conciliation Service. W. H. Rodgers, the Department's representative, was also unsuccessful in bringing the operators to the negotiating table.[535]

At this point the union leaders began to realize the operators would never recognize the union as the exclusive bargaining agent of the mine and mill workers. After meeting with local union leadership, a vote to strike was proposed and passed by 86% of the 700 voting.[536] But those members approving the strike represented less than 15% of the 4,800 men working in the mines and mills, therefore a number of those voting for the strike were most likely unemployed members. The strike, set to begin at midnight, May 8, 1935, was preceded that evening by the arrival of 20,000 rounds of ammunition, an omen of the violence to come.[537]

Many miners were unaware of the strike and showed up for work on the morning of May 9 only to find armed union members who, by force and threat, prevented the mines from opening.[538] Strikers blocked Highway 6 leading into Eagle Picher's Central Mill at Carden. Ottawa County Sheriff Eli Dry was unable to disperse the strikers.[539]

The miners had arrived at that fateful day of May 8, 1935, as a result of the operators' unwillingness to negotiate with the unions and because of the union's strategic and tactical errors. Could the strike have been avoided? It is conceivable that the management of one or more of the major mining companies could have had the foresight to have seen the outcome of the coming labor strife and to meet with union representatives. Given the district's history of being infertile ground for unionism, a few concessions would have gone a long way in appeasing and meeting the needs of miners. The operators were not callous toward the plight of the district's miners and had previously demonstrated a measure of concern and care through the establishment of clinics and strong support of the various charitable organizations. Without the tacit support of the major operators, small operators like Mike Evans would have much less power and would have followed the lead of the majors in negotiating with the unions.

The union also made crippling errors in both their unionization and strike efforts. First, the International Union's selection of Roy Brady as its organizer ultimately damaged the union's credibility through his false promises and theft of union funds. Second, the union had not arranged for adequate financial resources to back the strike, and the economic circumstances of the district were particularly desperate after five years of the Depression. Third, the proposed strike was ill timed because the operators were prepared for the traditional spring shutdown. Even without the strength of the soon to occur back to work movement, these errors made a successful strike difficult to achieve. If one or more of these errors could have been avoided, the operators may have possibly given more consideration to negotiations with the local unions before a strike was called. Nevertheless, action and reaction replaced thought and deliberation. The two opposing forces charged at each other as two trains on a single track, and the resulting collision would be disastrous for the district's miners and their families.

The "what might have beens" were of no consolation to Thomas Brown. He was in a difficult, no-win situation, and action had to be taken or the union would disintegrate. The mine owners were the only ones with any flexibility in their position. However, sensing victory and possible elimination of Brown's union, the operators chose to fight. This fight quickly took the form of encouraging the back-to-work movement started by the miners on May 16, only eight days after the strike was called. On that date Tom Armer, a blacksmith for one of the mines, met with a group of other men wanting to resume their employment and began the effort to break the strike. Discussions with several mine operators by some of Armer's group had already been held to determine the operators' willingness to reopen the mines. It was reported that operators were now willing to open their mines, and as a result, a meeting the next day (May 17) was scheduled at the Buffalo School west of Cardin to determine ways to end the strike and return to work. News of the meeting spread and several hundred men gathered at the school, both back-to-work supporters and those supporting the strike. A third meeting was held on May 18 where 1,500 men gathered to hear speakers for both sides. A fourth meeting was held on May 19 at the Miami fairgrounds to select a committee to meet with the operators to end the strike. Twenty-five hundred men showed up, and strike supporters were prohibited from speaking. At that meeting it was announced that a new organization would be formed to counter the International Union of Mine, Mill, and Smelter Workers. An almost unanimous show of hands indicated that the men present would join such an organization. Additional signatures brought to almost 1,800 the number of men who had signed petitions between May 16 and May 19 to request operators to resume mining. A fifth meeting was called at noon on May 20 at Baxter Springs to hear the report of a committee, with one representative from each company, appointed to plan a course of action. Brown and a significant number of pro-strike union members were among the 3,000 that came to the city park. Armer and the back-to-work speakers could not be heard over boos and catcalls from strike sympathizers. Likewise, when Brown attempted to address the meeting, he too could not be heard because of the noise created by the back-to-work supporters.[540]

Although pro-strike supporters were successful in disrupting and ending the back-to-work movement's meeting at Baxter Springs, it should have been obvious that the massive support for the movement's goals would make the success of the strike exceedingly difficult if not impossible. Clearly, the strike was in danger of imminent failure along with the destruction of the locals. To stem the successes of the back-to-work movement, the union demonstrated throughout the district the following day (May 21). The demonstrations ended with a march from Picher and Cardin to the Miami fairgrounds. Over a thousand men and women listened to Brown and other union spokesmen exhort supporters to stand firm. Two thousand would gather that evening for a rally at Picher. The crowd was so large that the meeting was moved from the union hall to a vacant lot on Third Street across from the city hall.[541] But the back-to-work supporters continued their organizational efforts and gathered additional signatures on back-to-work petitions.[542] There was little left for the strike leaders to do but continue the strike and encourage their supporters.

On May 25, only four days after Brown and the pro-strike miners attempted to stall the back-to-work movement at Baxter Springs, Armer and his committee held a strategy session at the John L. Mine operated by Mike Evans. Not intent on just ending the strike and getting the men back to work, the committee now wanted to start a new workers' organization. The committee of thirty contained a number of supervisors from the mining companies being struck. At this meeting Evans replaced Armer as the leader of the back-to-work movement.[543] Two rallies were held that day at the Miami fairgrounds, one at 10:00 a.m. with Mike Evans as principal speaker and a second at 2:00 p.m. .[544] The strike had begun on May 8. The back-to-work movement was born on May 16. By May 25, the movement had evolved into a quest not only to end the strike but also to establish a new union, now headed by a mine operator and a leadership group that included many mine supervisory personnel.

Evans was the same man discussed in a previous chapter who had served a prison sentence for making whiskey in an abandoned drift he owned, one of the largest stills ever found by federal authorities. Evans was a relatively small operator employing thirty-five men at the Craig Mine, but he had other entrepreneurial interests such the Connell Hotel in Picher that became the headquarters of the back-to-work movement. Other ventures included ownership of a nightclub and poolroom, part owner of the Picher Ford dealership, partnership in a business that sold used cable, in addition to having several real estate holdings.[545]

Mike Evans was to become the point man for the mine operators such as Eagle Picher. Although operating only a relatively small mine, Evans had lucrative contracts with Eagle Picher to supply ore to the Central Mill. He was a short, pudgy man with thinning hair. A born leader, tough, articulate, intelligent, persuasive, and experienced in managing men, he commanded intense loyalty from his friends and employees. Self-confident and a great delegator of authority, he could be a ruthless adversary when necessary.[546] Evans emerged from the May 25 back-to-work strategy meeting as the president of the newly formed Tri-State Metal, Mine, and Smelter Workers' Union that by the end of the day claimed a thousand men enrolled as members.[547]

On May 27, Sheriff Eli Dry was called to the Connell Hotel in Picher to escort Evans and other members of the new union to a large mass meeting at Miami. A large contingent of pro-strike supporters had gathered outside the hotel to block Evans' departure. The officers escorted the men to a pickup that managed to escape the scene as pro-strikers tried to stop it. Someone tossed a brick, and Sheriff Dry fired a warning shot. The mob surrounded the officers and severely beat them with pick handles and lead pipes.[548] The injured men were taken into the hotel. Thirty minutes later Dave McConnell, superintendent of the Oklahoma State Bureau of Investigation, and three carloads of heavily armed state troopers arrived and took

Dry and the other injured men to the Miami Baptist Hospital.[549] Dry was hospitalized for several days as a result of his injuries. The incident was particularly damaging to the striking miners' union. First, the union members started the violence contrary to instructions by union officials. Second, the assault on Dry and his officers thereafter effectively pushed them into support of the back-to-work movement.[550]

Joe Nolan, a hard rock miner who came from the lead mines of eastern Missouri to the Tri-State during World War I, eventually became the operator of the Luck O. K. Mine following a stint as Picher's police chief. The six-foot tall, 225-pound man was an imposing and fearless leader.[551] Word of the strikers' attempts to stop Evans and his men from leaving Picher and of the beatings of Dry and his officers reached the waiting crowd of back-to-work supporters at the Miami fairgrounds. Nolan gave a short, impassioned speech and exhorted them to follow him to Picher to confront the strikers "…and take back something that belongs to us."[552]

Many of those gathered at the Miami Fairgrounds were already armed with short pieces of steel, clubs, and guns. Nolan and the crowd that followed from Miami arrived at the Picher High School where a load of pick handles was distributed. Three to four thousand back-to-work supporters assembled in a long column, pick handles in hand, and marched four abreast through town.[553] Brandishing the pick handles and firing guns into the air, the back-to-work supporters passed the strike headquarters and drove the strikers from the street. A group of strikers sought refuge in their union hall. When violence between the groups appeared imminent, McConnell and his troopers intervened, and tear gas was used to disperse the mob surrounding the strike headquarters.[554] By nightfall, Governor Marland had ordered Colonel Ewell Head to Picher to command 120 Oklahoma National Guardsmen comprised of a rifle company from Wagoner and a machine gun company from Eufaula.[555]

On May 28, Evans met with the mine operators who, after hearing Evans' case for the new organization, agreed to open the mines. On May 29, loading began at one of the mines under the watchful protection of Colonel Head and his National Guardsmen. Fifteen guardsmen from the machine gun detachment were positioned in and around Mike Evans' Connell Hotel, the headquarters of the new back-to-work union where new members would sign two forms. The first was a form notifying the striker's union of the man's intent to resign, and the second was a blue card signifying that the man was a certified member of the back-to-work union, or Blue Card Union as it would become known.[556]

By the end of the first week in June, twenty-one operators had signed closed shop agreements with Evans' Tri-State Mine, Mill, and Smelter Workers' Union, the official name of the Blue Card Union. For all practical purposes, the strike lived on only in the minds of the thousand men who stubbornly remained members of the American Federation of Labor's International Union of Mine Mill and Smelter Workers.[557] Less than two months after the strike began, thousands of men who formerly were members of the International Union were now back at work as members of the Blue Card Union. The mines and mills were operating again, and public sentiment was against if not openly hostile to the strikers and their union. However jubilant as the Blue Card Union officials and their supporters were over the course of events, several were concerned about events transpiring in the nation's capitol.

May 27 was a disastrous day for the pro-strike union forces in the Tri-State because of the violence instigated by the union. It also proved to be a disastrous day for the labor movement in Washington, D.C. Section 7(a) of the National Industrial Recovery Act of 1933 provided federal recognition of labor's right to organize and pursue collective bargaining. Following arguments on May 2 and 3, 1935, before the United States Supreme Court, the court rendered its decision on May 27. The Act was found to be unconstitutional insofar as it permitted Congress to delegate legislative authority to the president to establish codes of fair competition. Additionally, the law was found to be unconstitutional in that it exceeded federal authority over intrastate commerce. As a result, the labor provisions of the Act were abolished.[558] Roosevelt and Congress quickly incorporated the Act's labor provisions into the National Labor Relations Act of 1935, known as the Wagner Act. The legislation became effective on July 5, 1935, less than two months after the Tri-State strike called on May 8.[559]

The National Labor Relations Board was created to promulgate rules and regulations with regard to the act and to initiate enforcement actions where necessary.[560] Mine operators had insisted the Wagner Act was unconstitutional, but passage of the Act had given the die-hard strikers some hope in the late summer and fall of 1935 that they would eventually prevail. However, a Kansas City federal judge delivered a lump of coal in the strikers' Christmas stockings by ruling on December 23 that the new law was unconstitutional. Although the judge's ruling was limited in scope and effect, many strikers gave up the fight and drifted away to seek jobs and a new life elsewhere.[561]

In spite of this and other court reversals, the heads of four of the International Union's locals in the Tri-State filed complaints of unfair labor practices against Eagle Picher Mining and Smelting Company on March 25, 1936, and against its parent company, Eagle Picher Lead Company, on April 23, 1936. Hearings were scheduled for May 25 at the National Labor Relations Board's regional office in Kansas City. Company lawyers obtained a restraining order from the Federal District Court (Northern District, Oklahoma) to block the hearings. The court subsequently issued a temporary injunction to stop the hearings into the International Union's complaints.[562]

Labor union gains in 1935 on the legislative front were somewhat weakened by a significant rife within the leadership of the American Federation of Labor. John L. Lewis headed the International Union of Mine, Mill, and Smelter Workers with which the Tri-State locals were affiliated. Lewis' union was affiliated with the American Federation of Labor. But following its 1935 convention, the AFL continued its policy of favoring craft unions over the industrial unions. In retaliation, Lewis withdrew his International Union and formed the Committee for Industrial Organization, and the International Union's striking locals moved their affiliation from the AFL to Lewis' new CIO.[563] The two labor organizations would bitterly oppose each other in the coming years and seek to enlist various unions into their respective camps.

The United States Supreme Court's ruling that the National Labor Relations (Wagner) Act was constitutional came on April 12, 1937, less than three months before the second anniversary of the founding of the Blue Card Union. In N.L.R.B. v. Jones & Laughlin Steel Corporation, the Supreme Court reversed a Circuit Court of Appeals ruling that the National Labor Relations Act was invalid as applied.[564] In spite of their insistence that the law was unconstitutional, the early worries of the Blue Card Union leadership with regard to the National Labor Relations Act of 1935 proved to be well founded. The Act specifically prohibited company unions, and with the Act's constitutionality now confirmed by the nation's highest court, the Blue Card Union faced federal intervention into its affairs and actions.

For two years the Blue Card Record (formerly the Metal, Mine, and Smelter Worker) had assailed the American Federation of Labor and its leadership with charges such as, "The A. F. of L. is infested with reds and communists," and called their leadership "dirty blood-sucking leeches" and "racketeers."[565] However, following the Supreme Court's decision on upholding the constitutionality of the National Labor Relations Act, the Blue Card Union's leadership attempted to lend an air of legitimacy to their company union and sought membership in the AFL. In spite of the previous vicious attacks on the AFL by the leadership in the Blue Card Record, the AFL leadership was anxious to recruit the Blue Card Union in its battle with the CIO whose International Union locals were still officially striking the mines in the Tri-State.[566] On May 23, 1937, any hopes held by the Blue Card Union leadership that affiliation with the AFL would forestall action by the National Labor Relations Board were dashed as the Tenth Circuit Court of Appeals lifted the injunction that prohibited hearings of the International Union locals' complaints against Eagle Picher filed in March and April of 1936.[567]

The complaints were significantly broadened to include charges of intimidation and violence by the Blue Card Union against International Union members, refusal of the companies to hire International Union members, active leadership by company supervisors in the Blue Card Union, the financial support by companies of the Blue Card Union newspaper, and company control of the Blue Card Union.[568] The hearing began on November 29, 1937, and ended five months later on April 29, 1938. The trial examiner confirmed most of the complaints of discrimination and violation of rights protected by the federal law. After review, the National Labor Relations Board sustained the examiner's findings, and the decision and order were issued on October 27, 1939. Back pay and reinstatement to jobs were ordered for approximately 200 claimants. Eagle Picher appealed the ruling on November 9, 1939, but the Eighth Circuit Court of Appeals agreed with the Board's order and issued an enforcement decree on May 21, 1941. Additional legal wrangling with regard to the formula for calculation of back pay would eventually be resolved when the United States Supreme Court dismissed arguments for using a formula that would have resulted in larger payments for back pay. The court's ruling was issued on May 28, 1945, ten years and twenty days after the beginning of the International Union's strike of the Tri-State mining fields. Final payments to the claimants were made in 1946.[569] It was a bitter victory for the few miners who received back pay and job reinstatement.

Troops lying in trenches and facing an enemy near at hand have little concern with the generals, their maps, and grand military strategies. Likewise, the complaints, court challenges, temporary injunctions, and political debate in Washington, D.C., were of little concern to the idle strikers and their hungry families. The strikers and working Blue Card members backed by the companies faced each other not across barbed wire but picket lines. Their warfare would be marked by two major waves of violence in the district—the first wave in the last half of 1935 and the second in the spring of 1937. The period would be called one of the bloodiest chapters in the labor history of America.[570]

Mention has been made of the violence at the beginning of the strike, including the beating of the Ottawa County Sheriff and his men on May 27, 1935. Apart from threats and intimidation, only two incidents of violence occurred prior to the assault on Sheriff Dry and his men. The first occurred on May 9 when picketers attacked Herschell Northcutt, a worker at the Ballard Mine.[571] The second occurred on May 20 when vocal anti-union activist Tom Wilkerson, operator of the Grace Walker tailing mill near Quapaw, was severely beaten by unknown parties while in Baxter Springs.[572]

Following the arrival of the Oklahoma National Guard the night of May 27, most of the violence in the Oklahoma portion of the district was attributed to the squad car forces that operated in the district with the consent and cooperation of both Sheriff Dry and Colonel Head, commander of the Guard. Formed shortly after May 27 by the new Tri-State (Blue Card) Union, between seventy-five and one hundred unemployed miners were deputized with the stated purpose of patrolling and protecting lives and property in the district. However, the money to operate the squad car units came from the companies and was merely funneled through the new back-to-work union.[573]

Squad cars containing four men would conduct nightly patrols of the Oklahoma portion of the district. These operations would be a major tool used by the Tri-State Union and the companies for intimidation and punishment of strikers and their supporters. Tactics included stopping and searching autos thought to contain striking International Union members. Strikers were beaten if found away from their homes at night. Many strikers were taken to the Tri-State Union headquarters at the Connell Hotel for interrogation by Evans and others. Ray Keller, a constable at nearby Hockerville and strike supporter, was forcibly taken from his home and transported to the Connell Hotel for questioning about incidents involving the strike. This occurred at the same time the hotel was being guarded by the Oklahoma National Guardsmen. Only the arrival of Cherokee County Sheriff's deputies while Keller was being taken from the hotel prevented Keller from being beaten.[574]

Two notorious criminals, Sylvester Walters and Roy Jamison,

were paid by the Tri-State Union to guard Evans in 1935. Jamison was involved in the beating of a sixty-six-year-old picketer. As the elderly man sat on a boulder, he was beaten with blackjacks, kicked, had his arm broken, and as one witness reported, they "tore the side of his face loose." Such violence was not caused only by criminals but also the roving squad car goons employed by the Tri-State Union. During June or July 1935, an elderly former justice of the peace was addressing a group of men, women, and children about constitutional rights as he sat in a vacant lot in Hockerville. Sheriff Dry and a deputy drove up and ordered the crowd to disburse. When the elderly man asked the crowd to remain, Sheriff Dry threw tear gas into their midst, causing injury to a number of women and children.[575]

The Cherokee County, Kansas, portion of the district had been peaceful until the June 7 opening of the Beck No. 3 Tailings Mill near Baxter Springs. The opening attracted fifty International Union strikers who threatened drivers delivering tailings for processing. Threats, road blocks, attempts to overturn trucks entering the facility, and refusal to disburse caused local officials to request that Kansas Governor Alf Landon send Kansas National Guardsmen to the county.[576] Colonel Charles Browne and 230 Kansas National Guardsmen under his command left their home stations after darkness on Friday evening and arrived in Cherokee County before daylight.[577] Violence erupted just before their arrival as three Empire District Electric Company towers were dynamited, plunging much of the area into darkness and cutting off power to many mines and mills. Other unexploded charges were found. Another incident involved the shooting of a union strike leader who was wounded in his legs and right arm while picketing near Treece.[578]

Following the Kansas Guardsmen's arrival, a steady reduction of troops commenced, and in less than three weeks from its beginning, the military occupation ended on June 27. However, the departure of the Kansas National Guard was ill timed given the Eagle Picher announcement that it would open its giant Galena Smelter the next day. Perhaps emboldened by the absence of the guardsmen, several hundred strikers assembled during the night across the road from the smelter. By morning strikers lined both sides of Highway 66 in an attempt to keep the smelter closed. Strikebreakers arriving from Missouri had their cars clubbed; windows broken, stoned, overturned; and were shot at. Fights erupted, and one person was shot. The strikers were able to prevent anyone from entering the mill. However, twenty-seven armed guards remained inside the smelter. A two-hour gun battle erupted at 9:30 p.m. between the guards and strikers. In response to an urgent request from county officials, Governor Landon reluctantly ordered the Kansas National Guard back to Cherokee County. Arriving in the early morning hours of June 29, Colonel Browne rescued the guards at the smelter and imposed martial law. Browne's men confiscated large numbers of weapons. This was followed by restrictions on the freedom of the press, assemblage, and speech. Arrests were made for rioting, perjury, and seditious speech. Additionally, Browne used his military powers to allow some of the mines to reopen as well as the Eagle Picher Smelter on July 16. By July 21, all guardsmen were removed with the exception of a major left in command of the county until August 7 when martial law was ended.[579]

Without question, the 1935 role of the Oklahoma and Kansas National Guardsmen's efforts to restore order also had the effect of breaking the strike. The new company sponsored back-to-work Tri-State Union enrolled four thousand men in less than ten days and signed agreements with thirty-three operating companies to return to work.[580] The International Union's strike would continue in name only over the next two years.

In the spring of 1937, Reid Robinson, now president of the International Union, began a new drive to organize the Tri-State mining district. Plans were laid over a several week period prior to the announcement of an organizational meeting to be held on Sunday, April 11 in the Blue Card Union stronghold of Picher. International Union supporters attempted to distribute circulars but were forced to flee or were beaten. In an attempt to thwart the International Union's efforts, Tri-State Union leaders met with 400 ground bosses and foremen on Saturday, April 10, and directed them to assemble their men the next day to stop the CIO meeting. Food and plenty of alcohol were promised, thus Blue Card Union members began assembling that evening. Ray Keller, the Hockerville constable who had been abducted from his home in 1935 and narrowly escaped a beating, was in Picher and observed consumption of alcohol in violation of the law. Keller took his concerns about the impending violence to the Picher police chief's office, only to find a number of Blue Card Union men with pick handles. Keller then called Ottawa Sheriff Walter Young and requested help in controlling the volatile situation. Young refused. Reid Robinson also called Sheriff Young, who again refused to come to Picher claiming that he had visited the city and found it quiet and peaceful.[581]

By Sunday morning, April 11, the situation in Picher had become explosive. A group of Blue Card Union supporters drove to Hockerville, where they seized and beat Constable Keller. Keller ignored their orders to leave town. The union halls of three International Union locals were ransacked, windows broken, furniture overturned, and records scattered, destroyed, or stolen. Street fighting, beatings with pick handles, and general mayhem were the rule of the day in Picher, and Evans and his Tri-State Union members were the rulers. Picher Police Chief Maness took no action to stop an attack by a mob of Tri-State Union members on a group of CIO supporters in Picher for the meeting. Robinson assessed the gravity of the situation and wisely cancelled the meeting scheduled for 1:00 p.m. . Fifteen men were hospitalized with injuries sustained from beatings. Sheriff Young would not arrive until late in the afternoon, long after the violence had occurred.[582]

Roving bands of Blue Card members drove to Treece, a mile north and just across the Oklahoma-Kansas line, and wrecked an International Union local's hall and stole its records. The CIO's International Union members were beaten, their union buttons forcibly taken and surrendered to the Tri-State Union in exchange for a bounty of one dollar paid for each button.[583] A rumored meeting of the CIO in Galena was cause for a second mass exodus from

Picher. Escorted by Sheriff Young to the state line at the northern end of Picher, one hundred cars and trucks carrying 500 men drove the eighteen miles between Picher and Galena. Assembling at their union hall, the Tri-State Union men marched to the International Union hall four blocks north on Main. Fifteen CIO men standing outside their International Union hall retreated inside.[584]

As Tri-State Union men neared the front of the International Union hall, Lavoice Miller ran to the front and began breaking the hall's plate glass windows with his pick handle. From inside the hall a shotgun was fired at the marchers, and a heavy exchange of gunfire followed. Smoke bombs were hurled at the International Union Hall, and the smoke and resulting pandemonium caused a theater across the street to empty. Nearby pedestrians fled the area for fear of being shot. When the smoke cleared, nine people had been shot, including a teenage boy. Lavoice Miller died eleven days later of complications resulting from his wounds. Many of the Tri-State Union men raced back to their union hall to gather guns and dynamite and call for reinforcements from Picher.

The International Union men inside the hall escaped out the back. Cherokee County Sheriff Fred Simkin had been in Treece investigating the attack on the International Union hall earlier in the day. He arrived back in Galena to find two Tri-State Union men looting the records of the International Union in their now abandoned hall. The sheriff arrested the men and then went to the Tri-State Union hall where he confiscated a number of shotguns and ammunition, ordered its occupants to return to Picher, and closed and padlocked the hall. As a result, a second assault on the International Union's hall and further bloodshed were averted.[585]

The growth in power and influence of the Tri-State Union had increased substantially in the nearly two years since its formation on May 27, 1935. The extent of that power as evidenced by the explicit support of many law enforcement officials in the district reached its peak on that violent and bloody Sunday of April 11, 1937. Ironically, the event leading to the breaking of the company union's stranglehold on the district would occur the very next day when the United

States Supreme Court handed down its landmark decision upholding the constitutionality of the National Labor Relations Act. The Tri-State Union's two-year rise to its pinnacle of power would be followed by a quick descent over the next two years. By 1939, following the loss of their appeal of the Board's ruling that the union was in fact a company union in violation of the National Labor Relations Act of 1935, the union ceased to have any influence on the affairs of miners as it related to labor and management relations.[586] However, the collapse of the Blue Card Union did not result in a rush to join the International Union. Bitter memories of the previous four years' fight against the International Union would keep many men from joining. Additionally, the miners knew the strong feelings of operators regarding membership in the union. This aversion to joining the union was substantiated by United States Bureau of Labor Statistics that only 23% of Tri-State miners were unionized in 1943 as compared to 53% in all of the mines and mills surveyed in the nonferrous metals industry.[587]

By the 1950s, the mines were rapidly being depleted as cheaper foreign imports drove down lead and zinc prices. The end of the era was near, and there was nothing a union or a company could do to bring back the glory, wealth, and jobs supplied by the one-time largest lead and zinc-producing region in the world. The area would continue to be described in superlatives, but not of the desirable kind. The lead and zinc gouged from the earth to help build a nation and fight its wars would leave a legacy that may last far longer than the memories of the mining era in the first half of the twentieth century—the legacy of being the nation's oldest and largest environmental Superfund site listed on the EPA's National Priorities List (NPL).

PART III—MINING AFTERMATH AND THE
QUAPAWS IN THE TWENTY-FIRST CENTURY

The Tri-State lead and zinc mining industry died in the 1950s with the last mining completed in the late 1960s. The remains of the mining operations became the United States' oldest and largest Environmental Protection Agency Superfund site. "The filthiest town know this side of Hell" describes the magnitude of the environmental damage and governmental efforts to address the problem. In addition to the environmental damage, the human cost to the Quapaws is examined. "The Quapaw—Into the Twenty-First Century" picks up the story of the Quapaw during the mining era, their great wealth and loss thereof, and their status in the twenty-first century.

"The filthiest town known this side of Hell"[588]

"Behold, your house is left unto you desolate."[589]

When businesses invest in the tools of commerce, the accountants periodically write off (expense) a portion of the cost over the life expectancy of the plant, equipment, and machinery. This periodic write-off is a cost of doing business and is called *depreciation*. Typically, a fully depreciated asset is at or near the end of its useful life and must be replaced. If businesses invest capital to purchase and develop properties containing a finite amount of natural resources, once again the accountants periodically write off a portion of the cost over the estimated period of time the resource will be produced. The periodic cost of producing the resource is an expense called *depletion*. Unlike depreciated property where worn out equipment is removed and replaced by new, once a natural resource is depleted, the business owner must move to other natural resource bearing properties or go into another business. Certainly, the mining fields of the Tri-State were of the depletable category.

The reality that the rich mining fields of the Miami-Picher district would not last forever seemed of little concern to the operators, miners, and residents of Ottawa County during the boom years of 1915 through the late 1920s. By 1921, two-thirds of the nation's zinc needs came from the Tri-State. The district was still producing a third of the nation's zinc and one-tenth of its lead in 1941. Yet in the early 1950s, the district would begin its death throes that would last for fifteen years. The district's slide into oblivion as the lead and zinc deposits were exhausted was hastened by the importation of foreign ores at prices below what it cost to produce the ore domestically. At the close of 1953, one industry representative testified about the plight of the domestic lead and zinc industry before the tariff commission in Washington, D.C. The industry's difficulties were illustrated by this testimony that revealed the number of operating mines in the Tri-State had declined from 150 in early 1952 to approximately 20 in

November 1953. In spite of industry pleas, the importation of foreign ores continued unabatedly.[590]

A little over three years later, the giant Eagle-Picher Company discontinued all operations of its mines and mills in the Tri-State. The announcement came on April 7, 1957. A *Miami* (Oklahoma) *Daily News-Record* headline stated, "Picher Area Shocked by E-P News: Other firms to continue in the Field."[591] By the end of that month, 1,100 Eagle-Picher employees were laid off, 500 in the Picher mine field and 600 at its Henryetta smelter. The shutdown included company's Central Mill and caused the loss of jobs for an additional 300 men in smaller mining companies that shipped ore to the mill.[592] In 1959, estimates of ore that remained in the district ranged between twenty and thirty million tons but were not economically producible until prices returned to "normal."[593]

Burial usually follows death. By 1969, the Tri-State was dead, but its body was too big and too expensive to bury. It would lay sprawled over forty plus square miles of Oklahoma and Kansas prairie, decaying and putrefying under the sun as the residents continued to live their lives among its bones and rotting flesh; themselves breathing, drinking, and wallowing in the corpse that surrounded them. Were the residents of Ottawa and Cherokee counties oblivious to these dangerous and unhealthy conditions? Any answer other than a resounding "no" would stretch credulity. They were the descendants of miners, and their families that had lived in the district for the better part of a century. Many worked in the mines or had fathers, grandfathers, and perhaps even great grandfathers who had fought the monster and died in mine accidents or suffered a slower death due to miners' con. The dangers of life in the district were well known from its earliest days. The end of mining would not lessen those dangers to the health and well-being of those living in and around the district's now silent mines and mills. It made it worse.

The magnitude of the problem with the Tri-State district's corpse was beginning to be recognized in the early 1980s. An editorial writer of the time quoted the Tar Creek Task Force charged with studying pollution of Tar Creek and area water wells. The writer stated that the

Task Force "'…predicted we would face zinc-polluted water for thirty-six years and that it would cost in the millions to construct treatment facilities…The mine water pollution of Tar Creek, area water wells and, fearfully, in the future of Neosho River and Grand Lake, has to rank as No. 1 problem facing us in Northeast Oklahoma.'"[594] The study to which the editorial writer referred was produced by the Tar Creek Task Force, created in 1980 by Oklahoma Governor George Nigh to investigate Tar Creek's acid mine drainage.[595] But the federal government would soon lift that responsibility from the state.

The Comprehensive Environmental Response, Compensation, and Liability Act of 1980 (know as CERCLA or, more popularly, the Superfund) was enacted to achieve two purposes: to identify and clean up sites contaminated by hazardous substances and to assign costs of the cleanup to parties responsible for the contamination.[596] Administered by the Environmental Protection Agency (EPA), the worst sites are placed on a National Priorities List (NPL). On September 8, 1983, the initial NPL was finalized and listed the 406 worst sites in the nation with regard to a Hazard Ranking System (HRS) based on certain criteria for determining priorities. The sites were then ranked according to their HRS scores with the highest scoring sites placed at the top of the list. The ranked sites were further divided into groups of fifty.[597] The Oklahoma and Kansas portions of the Tri-State Mining District were treated separately and therefore commanded two spots in the top fifty worst sites at number 48 and 49 respectively. It should be noted a high HRS score does not mean that it necessarily ranks at the top with regard to magnitude of the environmental damage or the cost to clean up the contamination. However, the Tar Creek and Cherokee County sites not only have a top fifty HRS score,[598] but the area has been called "the largest Superfund site in the nation…site of one of the Federal Government's most massive cleanup efforts."[599] According to a May 2005 report by the U. S. Army Corps of Engineers, the Ottawa County area encompassing the five communities known as the Tar Creek Superfund site was number one on the NPL at its inception and remains the nation's largest Superfund site.[600]

Since the inception of the NPL, 1,547 sites have been listed with

1,239 sites presently remaining on the list of which 172 are federally owned sites. The 308 deletions were removed from the list because the EPA determined that no additional action was necessary to protect the environment and human health.[601]

Remediation or cleanup of a contaminated site requires a series of steps. First a site is investigated to determine the type and extent of the contamination. Next, a system is designed to accomplish the cleanup. The various elements of engineering and construction of a cleanup system include site preparation and the installation of equipment and machinery. Each of these phases requires adequate and consistent funding and the cooperation and coordination of a myriad of governmental and private interests.

Of the 1,239 sites currently on NPL, 960 have had the engineering and construction phases completed. However, the cleanup systems put in place may operate for years or decades before the sites are brought to acceptable environmental and health standards.[602] Many sites have multiple problems with which to deal. Tar Creek is a prime example with several major environmental hazards that must be cleaned up, and each of the multiple hazards will require a separate remediation plan.

Water was always the enemy of the miner. For eighty years miners of the Tri-State fought a battle with the natural springs flowing up from the Roubidoux as well as water from the Boone aquifer as the waters attempted to fill the mine caverns and drifts. Thirteen million gallons of water per day were pumped from the mines from depths ranging from 90 to 300 feet below the surface.[603] As the mines were closed beginning in the early 1950s until the last mine was shut down in 1969, the pumps that removed huge volumes of water from the mines were turned off. Without the pumps, the mines filled with water, and the sulfide and iron deposits in the mines caused the water to become acidic. To reopen the mines would present several insurmountable problems that would prevent such ventures. Any new pumps installed to pump out the acid water would quickly corrode and become inoperable. Also, acid water pumped to the surface

TAR CREEK

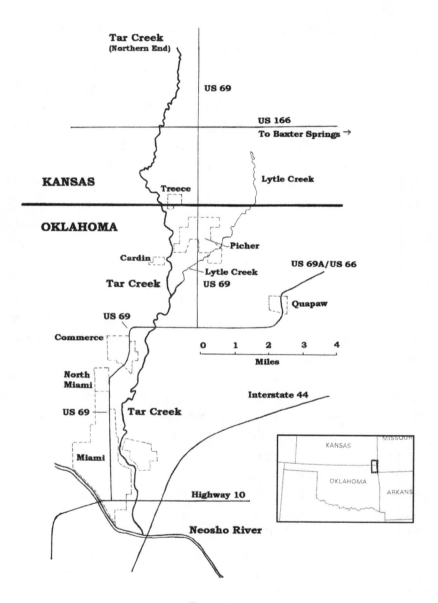

Figure 13

would pollute the surrounding streams and lakes and eventually flow southward and pollute even larger bodies of water.

But Mother Nature did not wait for new pumps to bring the acid mine water to the surface. In 1978, nine years after the last pump was shut down, the water began bubbling to the surface in a field near Commerce, but the water flowing from the mines was now acidic and contained significant amounts of heavy metals. Not only was the acid mine water polluting the surface, some of the mine water was finding its way downward to the Roubidoux aquifer. Several of the deep wells supplying area cities and towns became contaminated as acid mine waters corroded the well casings that pass through the Boone Formation to the Roubidoux aquifer below. As the corroded well casings allowed acid water to enter, the wells were abandoned but continued to provide a way for the acid mine water to migrate down to the Roubidoux aquifer.

It is estimated that there are seventy-six thousand acre feet of acid mine water in the caverns and drifts of the two Superfund sites.[604] If this estimate is correct, this amount of acid mine water is equivalent to 590 million barrels of polluted water (at forty-two gallons per barrel). Water continues to flow from the Roubidoux aquifer into the mines. The mines overflow into Tar Creek and other small creeks and tributaries that flow into the Neosho River and ultimately into the Grand Lake of the Cherokees, the huge Northeastern Oklahoma jewel completed in the 1940s when the U. S. Army Corps of Engineers dammed the Grand River formed by the confluence of the Neosho and Spring Rivers. The seventy-two square mile lake with 1,300 miles of shoreline[605] is at risk of contamination if the acid mine waters reaching the surface are not stopped. The problem of acid mine water and the resulting contamination of the surrounding rivers and lakes and the aquifers supplying area drinking water may prove to be the most expensive and difficult to solve of all of Tar Creek's environmental problems.

Tar Creek's second, most obvious, and eye-catching environmental and health hazards are the huge chat piles that dot the landscape. Approximately twenty major piles remain within a three-mile radius of Picher's center at Connell and A Streets. The mine tailings are the waste product generated when the lead and zinc ore was separated from the rock and other minerals during the milling process. Lead and zinc ores accounted for only 4 to 6% of the rock and ore brought to the surface. The tailings or unwanted chat accounted for the remaining 94 to 96%. These chat piles would be re-milled one or more times over the years as milling technology improved, but 10% of the lead and 25% of zinc remained in the chat piles. The magnitude of the chat problem is better understood when one realizes that 5,000 surface acres are covered with mine tailings in the form of chat piles, mill ponds, flotation ponds, chat-filled sinkholes, and fields leveled with an overlay of chat.[606] Much of Picher and Cardin are built on lands that once were the site of numerous chat piles. Over 800 of the 5,000 acres containing mine tailings were once wet or dry flotation ponds.[607]

Seventy-five million tons of chat remained on the surface of the ground.[608] It is just as remarkable that over one hundred million tons of ore and mine tailings have been removed from the Picher field assuming the field produced the estimated 181 million tons of ore and rock as has been reported.[609] The majority of the chat was hauled away by rail cars and used for ballast on track beds. The giant Central Mill tailings pile was called the largest in the world[610] but was completely removed over a period of years. Until a recent remediation effort by the EPA somewhat lessened the risk, blowing dust from the chat piles and tailings ponds have raised the lead levels in the blood of area residents with particularly devastating effect on the health and learning abilities of the district's young children.

The third major environmental and safety hazard of the Tar Creek site involves 300 miles of tunnels and over 1,300 mineshafts,[611] with approximately 450 of the shafts remaining open and often unfenced.[612] Additionally, the mining era saw 30,000 boreholes drilled in the Miami-Picher district.[613] There are a total of 2,500 acres of mines that underlie Ottawa County, some reaching depths of 380 feet. Some of the caverns are reported to be as tall as 125 feet and as wide as 1,000 feet.[614] The 2,500 plus acres of mines scattered under Ottawa County are equal to four square miles, nearly 10% of the forty-one square mile Picher Superfund site.

Cave-ins of abandoned mines, or subsidences as they are called, have occurred with regularity over the years. Children, hunters, hunting dogs, cattle, homes, autos, roads, businesses, and virtually any thing residing on the surface above the abandoned mines have been claimed by subsidences. Some have been spectacular in size. The largest reported collapsed surface area measured 560 feet long and 400 feet wide, an area just over four acres.[615] One collapse was so complete and uniform that cottonwood trees were left standing in the middle of the crater.[616] The Lost Trail Mine located at the south edge of Commerce was prophetically named. Highway 66 parallels the edge of a deep cave-in of the mine that caused the roadbed to sink on two different occasions.[617] In 1950, a four-block area straddling Picher's Main Street was fenced and abandoned after a nearby subsidence occurred. Over 200 home and business owners were ejected. The buildings were razed and the area fenced.[618] A 1967 subsidence engulfed three homes, and another family escaped with their lives as their home disappeared into a 1974 subsidence. A motorist was killed as he drove into a subsidence that had occurred in a roadway.[619] Recent borings by a team of subsidence experts have revealed the possibility of a catastrophic cave-in two miles south of Picher under a portion US 69, the heavily traveled major road in the Tar Creek Superfund site. It is believed that the roof of the mine under the roadway may be as close as thirty feet from the surface.[620] Although most cave-ins have occurred in less populated areas, three subsidences have occurred within 1,000 feet of Picher High School and Elementary School.[621] Most cave-ins occurred before 1952. By 1986, fifty-nine major subsidences had occurred that were ninety-five feet or greater in diameter. Half of these collapsed mines were associated with mineshafts. The remaining non-shaft collapses usually were the result of multiple mine levels, large cavern heights, weak roof rock, and support column removal by gougers at the end of the mining era.[622] Many additional subsidences have occurred since Luza's 1986 report. The 2000 Subsidence Committee Report listed seven with many more in remote areas not included that had occurred since 1986. The largest of the new subsidences occurred 300 feet behind the Commerce Police Station in 1994–1995 and measured fifty feet wide by seventy feet long with a depth of 140 feet.[623]

Subsidence problems of the mining district are not confined to Tar Creek. The Cherokee County, Kansas, Superfund site on the north side of the Oklahoma-Kansas border adjacent to the Tar Creek site also has experienced substantial cave-ins of the undermined area. A large sinkhole opened next to a hundred-year-old Galena building in August 2006 and has grown to approximately eight feet in diameter with a depth greater than sixty feet.[624]

Due to growing concerns with regard to threat of new subsidences and through the efforts of Oklahoma Senator James Inhofe, an eighteen-month U.S. Army Corps of Engineers' subsidence study was initiated and funded. The study published in 2006 covered 4,400 acres, almost seven square miles of the forty-plus square mile Tar Creek Superfund site. Of the 4,400 acres studied, 1,270 acres were found to be undermined including 34% of Picher, 29% of Cardin, and four and one-half miles of nineteen miles of major transportation corridors in the study area. Of the 1,270 undermined acres, 286 locations covering eighty-eight acres were determined to have a potential for subsidence from less than two feet to greater than fifty feet deep. Forty-seven of the locations had a potential for a subsidence from twenty-five to fifty feet deep. Twenty-three of the locations had subsidence potentials of greater than fifty feet. Five areas within Picher including the town's Reunion Park have a subsidence potential of greater than fifty feet.[625]

It has been over a quarter of a century since the creation of the environmental Superfund and the inclusion of Tar Creek in the initial NPL two years later. Given the considerable passage of time in relation to the small amount of progress in remediating the site, several questions arise. What has been accomplished? What remains to be done? What will the cleanup cost? And finally, why has it taken so long?

The easiest of the four questions to answer relates to what has been accomplished in the Tar Creek Superfund site. When Tar Creek was placed on the NPL, there was an imminent danger that acid mine waters would contaminate the deep wells supplying several cities and towns in the mining district. The rich veins of lead and zinc in the Tri-State district occur in the Boone formation, a layer of limestone and dolomite that may reach as deep as 400 feet below the surface. Below the Boone is the Roubidoux aquifer containing water-producing sands and inhabits several geologic strata with the Roubidoux formation being the principal aquifer for Ottawa County.[626] The Boone aquifer in the Boone and other formations above the Roubidoux aquifer is the primary source of water derived from private, shallow wells whereas the deeper Roubidoux aquifer supplies most of the cities, towns, and rural water districts in the mining district.[627]

One may ask why the acid mine water became a problem only after the cessation of mining in 1969. As mining progressed, the pyrite-rich sulfide wastes and surface areas in the mines were exposed to oxygen. The oxidized sulfides in mines easily dissolved as the water began filling the mines and therefore produced acid mine water. The acidic water then reacted with the rock in the walls, ceilings, and floors of the flooded mines. The metals in the rock dissolved when exposed to the acidic water and, consequently, the water that began seeping from the flooded mines was not only acidic but contained high concentrations of zinc, lead, cadmium, and iron.[628]

There are two ways for the acidic, contaminated water in the mines to reach the underlying Roubidoux aquifer. First, water (from the Boone aquifer including acid mine water) may naturally flow downward through the intervening strata. However, the strata between the Boone and the Roubidoux have been determined to have relatively low permeability, meaning that water will not easily pass through the formations. Additionally, EPA scientists have reported that it is unlikely that fractures in the 300 to 400 feet of strata between the two aquifers would be so interconnected as to permit seepage of contaminated Boone waters into the Roubidoux waters.[629] Scientists have projected that it will take 15,000 to 25,000 years for the district's mine water to naturally cross the distance from the Boone to the Roubidoux. Also, because of the magnitude of the upward flows of the Roubidoux waters to the Boone and to the surface thereafter, the mines are projected to have been flushed several times resulting in little long-term impact on the Roubidoux aquifer. Because of this flushing action, scientists project that it will take a relatively short time (60 to 100 years) for the acid mine water to be replaced with water that is relatively uncontaminated.[630]

The second way contaminated water may reach the Roubidoux is through water wells punched through the Boone formation to reach the Roubidoux. Casings of the deep Roubidoux wells would eventually corrode where they passed through the acidic Boone formation waters. The corrosive water would eat holes in the pipes and provide an avenue for the acid mine water from the higher Boone formation to flow down and contaminate the Roubidoux.[631] One of the first actions of the EPA following creation of the Tar Creek Superfund site was to plug 83 abandoned wells that passed through the Boone formation to the Roubidoux aquifer and to build dikes and diversion structures to prevent surface water from entering two abandoned mines. The plugging of the initial group of wells and dikes was completed by December 1986.[632] This project, called Operating Unit 1 (OU 1), was the first implemented in the Superfund site. Remediation projects within the Superfund site are assigned different operating unit numbers. The well plugging and monitoring actions that began with OU 1 are ongoing.

The second action (OU 2) began in 1997[633] and deals with remediation of residential yards and public areas in the five towns located in the Superfund site (Picher, Cardin, Quapaw, Commerce, and North Miami) with additional funds provided for community health education and blood screening. Almost 2,200 residential yards and public areas have had soil contaminated with chat and chat dust removed and replaced.[634] Although the residential yard and public areas soil remediation efforts are small in comparison to the massive surface contamination remaining in the district, it appears to have yielded a remarkable reduction in the blood lead levels (BLL) for the young children of Tar Creek. In October 2004, a report to Congress by the Agency for Toxic Substances and Disease Registry of the Center for Disease Control indicated that only 2.8% of children (7 of 250) between ages one and five living in the Tar Creek Superfund site in 2003 had a BLL in excess of ten micrograms per deciliter. This compares to a 1995 study in which 31.2% of children tested (67 of 215) in the same age range had a BLL of 10 micrograms per deciliter. One study of U.S. children in the same age range during 1999–2000 indicated that 2.2% had a BLL of 10 micrograms per deciliter or greater, slightly below the percentage in the 2003 Tar Creek study.[635]

The third EPA initiative (OU 3) arose as an emergency action with the discovery and removal of a number of drums and chemicals from the former Eagle-Picher Mining Laboratory located in Cardin.[636] The fourth operating unit (OU 4) is a remedial investigation and feasibility study that includes chat piles and flotation ponds, mine and mill residue. The fifth EPA action (OU 5) is a characterization of the surface water and sediment in the Spring and Neosho River basins.[637]

The EPA's work in Tar Creek has primarily concerned itself with emergency actions on matters requiring relatively quick response to address immediate threats to the environment and health. But these efforts are miniscule and piecemeal in comparison with the answer to our second question posed earlier regarding what work remains. The short answer is removal of huge chat piles, dealing with blowing chat dust, remediating contaminated mine water, and addressing the dangers posed by subsidences (actual or threatened), open shafts, and bore holes.

Our third question concerned the ultimate cost of cleaning up the Tar Creek Superfund site and how many years it will take. The answers cannot be given or even estimated until completion of extensive site assessments and development of remediation plans for each of the environmental and health risks. Approximately $110 million has been spent on the first three operating units. Another $45 million will be spent on the Oklahoma Plan discussed below.[638] And it is likely that hundreds of millions remain to be spent to complete the task over decades to come.

Our fourth and last question involves why there has been relatively little cleanup progress made in the twenty-plus years that have elapsed since Tar Creek was placed on the NPL. The answer is complex but generally revolves around several issues. First and probably the most important, the physical size of the cleanup site, the severity of the contamination, and the magnitude of the cleanup effort required makes Tar Creek unique compared to most if not all other major Superfund sites. It is easier to deal with sites that are of a manageable size and where demonstrable progress can be achieved. Apart from the magnitude of the problem, there are other reasons for the lack of progress in cleaning up Tar Creek. Inadequate or nonexistent remediation technology has prevented progress. How does one remediate 590 million barrels of acid mine water? With so many sites requiring funds, inadequate funding for Tar Creek has been a deterrent. Tar Creek could consume the entire EPA annual budgets for several years. The slowness with which progress has been made is also due in part because of the legal obstacles (including liability and ownership issues), competition between governmental agencies, and a lack of coordination of cleanup efforts between those agencies. The maze of major governmental bodies involved include the EPA and two of its districts, Oklahoma Department of Environmental Quality, Bureau of Indian Affairs, governments of various Indian nations, Department of the Interior, U.S. Army Corps of Engineers, U.S. Geological Survey, Bureau of Land Management, and various state and local environmental groups.

Not the least cause of inaction is the politicization of the various issues surrounding the Tar Creek cleanup. In May 2003, under the leadership of Oklahoma's senior U.S. Senator James Inhofe, leaders from the highest levels of federal, state, and tribal governments announced what would become known as the Oklahoma Plan for Tar Creek. The plan proposed to bring quick results where there were no legal obstacles preventing action and centered on division of the Superfund site into six sub-areas. The Picher-Cardin sub-area resides at the center with the remaining five sub-areas forming the perimeter. Under this plan the perimeter areas of the Super-fund site including North Miami, Commerce, and Quapaw would be addressed over three to five years through the achievement of four objectives: improving surface water quality, reducing exposure to lead dust, attenuating mine hazards, and land reclamation. A remediation plan and action in the Picher-Cardin sub-area would continue to be pursued as legal and ownership issues are resolved.[639] At the inception of the Oklahoma Plan, Senator Inhofe was the chairman of the U.S. Senate Committee on Environmental and Public Works. The Committee has jurisdiction for environmental matters over the EPA, Army Corps of Engineers, U.S. Department of the Interior, and the White House Council on Environmental Quality. From this powerful position Senator Inhofe coordinated the first Memorandum of Understanding signed by a number of key federal agencies involved with Tar Creek to gain cooperation in addressing the various issues and activities required for cleanup of the site.

To illustrate how the cleanup process can be politicized, one need only to look at the response to the Oklahoma Plan. The two principal sources of opposition to the plan came from Democratic Congressman Brad Carson (representing Oklahoma's second congressional district from 2000 to 2004) and the *Tulsa World*, the state's second largest newspaper. Historically, the newspaper has been a strong supporter of Democratic Party issues and, with infrequent exceptions, endorses most of the Democratic candidates for public office.

The possible retirement of Don Nickles, Oklahoma's senior United States Senator at the time, with the expiration of his term in late 2004 was the hot topic of conversation in Oklahoma political circles during the spring of 2003. The rumors proved true as Republican Nickles announced his retirement in early October 2003, and Democrat Carson almost simultaneously announced plans to run for the office to be relinquished by Nickles. However, in the spring of 2003, well before the October announcements, the political maneuverings had begun with the first rumblings of a possible retirement. Tar Creek would become one of the major issues in the

Democrat vs. Republican Senate race. Inhofe's efforts to get something done with Tar Creek and the resulting development of the Oklahoma Plan for Tar Creek left Democrats little wiggle room on the issue. However, Democrat Carson began beating the drum of quickly moving Picher-Cardin residents from their homes through a buyout arrangement before further environmental cleanup action was implemented through the Oklahoma Plan.

Although Inhofe was not Carson's political opponent in the upcoming elections, Carson's opposition to the Inhofe plan generated considerable controversy with a corresponding increase in headlines and exposure for Carson. In late April and early May, Carson began pushing his buyout idea.[640] On May 14, Inhofe announced the first phase of the Oklahoma Plan for Tar Creek.[641] Thus began a series of *Tulsa World* news articles, political cartoons, and editorials that featured the Tar Creek Superfund site. The magnitude of the newspaper's effort to generate a campaign issue and publicity for Carson may be seen when the newspaper's archives are examined. A website search of the newspaper's archives for the words "Tar Creek" revealed ninety-three stories or mentions in 2002. In 2003, there were 257 stories or mentions. Even more revealing is the substantial increase in number of stories or mentions beginning in May 2003 when the Carson buyout plan and Inhofe's Oklahoma Plan for Tar Creek were announced. Between January 1 and April 30, 2003, there were only thirty stories or mentions of "Tar Creek." Beginning May 1 and continuing for the remainder of the year, there were 227 stories or mentions of Tar Creek. The number of stories or mentions dropped to 164 in 2004 and 69 in 2005.[642] Whether one agrees with one side or the other, it is obvious that politics have played a significant role in generating controversy and impeding progress of the Tar Creek cleanup.

Most of the newspaper's articles played up the controversy and focused on those aspects that supported the buyout plan of Carson. Less than six weeks after the Oklahoma Plan's announcement, the newspaper editorialized "Inhofe plan not the answer." The editorial stated that, "Oklahoma's Sen. Jim Inhofe has outdone himself. His plan to move chat piles that threaten the health and futures of Picher and Cardin residents is filled with almost as many holes as the Tar Creek region."[643] The newspaper's call for quick action to get the people out rings hollow when one considers the fact that the people in Picher and Cardin are second, third, and fourth-generation residents who were well aware of the dangers of living in the district including exposure to the dust and heavy metals as well as cave-ins and open mine shafts. Many have lived in the area for decades and raised their families in the minefields. For the newspaper to call for stopping the environmental cleanup efforts designed over a period of years by a myriad of governmental agencies, scientists, technicians, and contractors was irresponsible. Additionally, the campaign to get the people moved from Picher and Cardin as well as the promotion of a limited buyout of residents' homes were most likely counterproductive. Many who contemplated moving before calls for a buyout may have decided to not move because of their hopes of receiving greater compensation for their properties than they otherwise would have received. In fairness to the newspaper, it must be noted that a former Republican governor had proposed buying out the residents of the site and turning the property into a giant wetland area. However, this proposal was made before a comprehensive remediation plan and schedule had been put in place as was provided by the Oklahoma Plan for Tar Creek. Little was made of this plan by the newspaper until a buyout plan became a politically expedient campaign issue for Carson.

Republican Tom Coburn, the second district's former Congressman, and the second district's incumbent Democrat Brad Carson each became their respective party's candidate for the November 2004 U.S. Senate general election in Oklahoma. Coburn won, and the State of Oklahoma sponsored a small-scale buyout in 2005 of properties of families with children age six or below. But the *Tulsa World's* opposition would continue. In early 2005 the newspaper continued to rail in one editorial, "...Don Nickles and Jim Inhofe dragged their feet. Nickles spent his entire senatorial career of 24 years without significant action on Tar Creek, and Inhofe, when he did become interested, argued for cleaning up the environment instead of moving people out of Picher."[644] This editorial conveniently ignores the fact that Democrats held the Second Congressional District and one of Oklahoma's two U.S. Senate seats during the first fourteen years of the Superfund. Republican James Inhofe captured the Democratic Senate seat in 1994. From 1994 until his self-imposed term limit in 2000, Republican Tom Coburn was the district's Congressional representative after which the Democrats regained the seat in 2000. Furthermore, the first significant environmental remediation in Tar Creek began during Coburn's six-year tenure in the House and after Senator Inhofe's election in 1994.

Ironically, the buyout pushed by the Carson and the newspaper became a reality for many Picher and Cardin residents for reasons other than political gain. Because of the imminent danger posed by the threatened subsidences as revealed by the U.S. Army Corps of Engineer's January 2006 study, Senator Inhofe caused a portion of the funds allocated for Tar Creek remediation under the Oklahoma Plan to be diverted for the buyout of homes of residents and businesses in the affected area.[645]

Following three years of editorials, political cartoons, and less-than-balanced reporting, all critical of Senator Inhofe's efforts on behalf of Tar Creek, a recent *Tulsa World* editorial appears to grudgingly give the senator credit when it stated, "Inhofe is the first federal Oklahoma politician who did more about Tar Creek than visit the place and tut-tut over the problems."[646]

Neither political party nor their supporters are exempt from political maneuverings that hinder responsible governmental action and advancement of the public's interests, and Tar Creek has certainly been the victim of some of these maneuverings. The Superfund site will continue to be the bone of contention in a test of wills of elected officials and bureaucrats alike. However, because of the efforts of Senator Inhofe and others, both Democrats and Republicans, the cleanup of Tar Creek is progressing faster than at any time in its Superfund history.

—

Some of the residents gladly accepted the government buyout. Others stated that they would never leave the town they loved. Although the Tri-State lead and zinc mining industry had been dead for many years, the town of Picher was still struggling to survive by spring 2008. Daytime temperatures begin edging closer to the nineties each day. It was a typical Saturday evening on May 10, 2008, and the clouds to the west promised another round of thunderstorms as was so often the case on the southern plains at that time of year. If there had been a doubt about Picher's future, the forces of nature would make the final decision on that fateful evening. The developing thunderstorms produced a tornado near Chetopa, Kansas, just north of the Oklahoma border and fifteen miles west of Picher. It swept across the prairie through the northern edge of Oklahoma and on to Granby, Missouri, thirty miles east of Picher. In its path lay Picher. The EF-4 tornado with winds between 166 and 200 miles per hour cut a half-mile wide path of destruction through the town killing six people. Over one hundred were injured. One hundred fourteen homes were destroyed and thirty others sustained major damage.[647] What decades of mining, environmental hazards, and government buyouts could not destroy, nature had accomplished in a matter of minutes. Picher was no more.

It would appear that much of the center of the Tar Creek Superfund site will again revert to a quiet prairie presided over by circling hawks gliding on thermal updrafts. The only sounds will be the wind whistling through the broken windows of abandoned buildings and perhaps the drone of remediation equipment, like that of a hospital respirator, performing its seemingly endless task of bringing life to the injured land. The chat piles will be hauled away or leveled and grasses will again cover the scarred landscape. Life will go on at its fringes in North Miami, Commerce, and Quapaw. But life as it was once known by the Quapaws, the miners, and residents of Picher and Cardin will be forever gone. In time, even the memories of what occurred beneath Tar Creek's grassy veneer will fade.

The Quapaw—Into the Twenty-first Century

"But he that shall endure unto the end, the same shall be saved."[648]

The Quapaws' struggle for survival lasted for centuries. Driven from the Ohio Valley by warring Indian tribes, their sojourn in their Arkansas homeland would last a little over 150 years before they would be moved to a reservation in northeastern Oklahoma. Now, almost 200 years later, a few of those Quapaws' descendants still reside on the tribal reservation. For a brief time in the early part of the twentieth century, it appeared the Quapaws' troubles had come to an end with the discovery of the great wealth that lay beneath their reservation. This is where our narrative of the Quapaws' history was interrupted as Indian Territory moved toward statehood.

The tribe accomplished two remarkable achievements just prior to the discovery of the large deposits of lead and zinc on their reservation—the adoption of other Indians into the Quapaw tribe and the self-allotment of their reservation to individual Quapaws. Both were firsts in the Indian world's dealings with the United States Government, and both would have dramatic consequences for the tribe in the twentieth century.

The Quapaws had received approval of their allotments (individual ownership) in March 1896. However, the Quapaws were still restricted from alienation (selling) their land without government approval, a restriction that would last twenty-five years. By the expiration of the restrictions in 1921, the Quapaws would supposedly have made the transition to a white-dominated society and end their ties to the government.[649] However, the discovery of lead and zinc on the reservation in 1897 and in vastly greater quantities in 1914 would complicate that transitional process.

Beginning in 1895, six months before the government's approval of the Quapaw allotments, adopted Quapaw A. W. Abrams and others had obtained numerous mineral leases from several allottees. The Indian office did not approve the leases and proceeded to shut down and evict the lease owners and their enterprises. Abrams and Samuel Crawford, no strangers to lobbying on Capitol Hill, traveled to Washington in November 1896 and effectively lobbied Congress to allow Quapaw Agency Indians to lease their lands.[650] On June 7, 1897, Congress passed the requested legislation:

> That the allottees of land within the limits of the Quapaw Agency, Indian Territory, are hereby authorized to lease their lands, or any part thereof, for a term not exceeding three years, for farming or grazing purposes, or ten years for mining or business purposes...Provided, That whenever it shall be made to appear to the Secretary of the Interior that, by reason of age or disability, any such allottee can not improve or manage his allotment properly and with benefit to himself, the same may be leased, in the discretion of the Secretary, upon such terms and conditions as shall be prescribed by him.[651]

The legislation gave the Quapaws a greater measure of freedom on the use of their land, but it also opened the doors for considerable fraud against the Quapaws from a host of unethical businessmen, swindlers, and confidence men who soon controlled most of the reservation through leasing arrangements. Much of the value of the mineral-rich land was diverted from the allottees through a variety of these lease arrangements.[652]

Many leases were written for excessively long terms, ninety-nine years in one instance. Overlapping leases were used to extend the term of a lease before the current lease expired and rarely contained any significant changes. Leases would be written without a requirement for capital expenditures by those responsible for developing the property, thereby tying up the allottee's land without a requirement to produce income and pay royalties. Another scheme involved pyramiding royalties whereby an allottee would lease his land for a royalty percentage, say 5%. The person leasing the land would not develop it but sublease it for a higher royalty percentage. The sublessor may again sublease the property. The final operator who eventually leased the property for development would pay a royalty as high as 25% with only 5%

going to the allotted owner of the property. Due to the operator's high overhead costs caused by the excessive royalty payment, the operator could only mine the most valuable and easily retrievable ore, leaving much of the ore in the ground and thereby shortening the mine's life and allottee's royalty. Other unscrupulous persons would purchase an allottee's future stream of royalties for an immediate but nominal amount of cash. Following a decision by a judge in 1897, restrictions against inalienability ended with the passing of an original allottee's property to an heir. As with leasing, there was a substantial amount of fraud involved in acquiring Quapaw land from the allottees' heirs.[653]

Given the significant inequities and fraud visited upon the Quapaws with regard to their land and mineral wealth that lay below, the Bureau of Indian Affairs issued regulations in January 1907 that brought the mining transactions of sixty-five Quapaw allottees under federal government supervision. As he did in 1895, Abrams went to Washington along with Samuel Crawford to have the regulation rescinded. Chief Peter Clabber, who had been included in the group to be supervised, and other Quapaws were indignant at being judged as needing a guardian and argued against the government's supervision. Following a trip to the reservation by the Secretary of the Interior's personal secretary to interview the Quapaws affected by the regulation, the secretary recommended that the rules be rescinded except for twenty-five Quapaws who should have a guardian appointed by the local courts. However, it would be November before the Interior Secretary would amend the regulation to just the twenty-five requiring government supervision. Efforts were made to appoint guardians for the twenty-five, but Abrams and his attorneys were successful in getting the court to not appoint guardians.[654]

Because of the weight of evidence of widespread fraud in securing leases and lands from the Quapaws, the Attorney General sent special assistant Paul Ewart to the mining district to exert pressure or take whatever legal action necessary to assure that all business dealings between outsiders and the Quapaws were legal. Over a fifteen-month period, Ewert was highly successful in securing the return of approximately nine thousand acres of mineral lands obtained by fraud and in canceling agricultural and mining leases on almost thirty thousand acres. As part of his efforts, Ewert instituted legal action against Abrams to annul a series of overlapping leases. Ewart lost his case against Abrams in district court and again upon appeal and ruling by the Circuit Court in 1912.[655]

The issue of overlapping leases and assignment of royalties would arise in another case and be fought all the way to the U.S. Supreme Court. The lawsuit was filed against Charles F. Noble, John M. Cooper, A. J. Thompson, A. S. Thompson, and V. E. Thompson regarding a series of overlapping leases and assignment of royalties by Chief

Peter Clabber and Tall Chief (Photo courtesy of Dobson Museum & Memorial Center, Miami, Oklahoma)

Quapaw Chief Victor Griffin and Vice President Charles Curtis (Herbert Hoover's vice president) (Photo courtesy of Baxter Heritage Center & Museum, Baxter Springs, Kansas)

Charley Quapaw. Quapaw originally leased his land on January 11, 1902, for a period of ten years for a sum of ten dollars and a 5% royalty. Thereafter, a series of five additional leases were signed by Charley Quapaw, each with basically the same 5% royalty and a payment ranging between ten and twenty-five dollars. The last lease was signed on July 28, 1906, and extended the lease for twenty-five years for twenty-one dollars and the same royalties and minimum rental. In addition to the leases, Charley Quapaw had assigned a portion of his royalties. In the proceedings, Charley Quapaw was found to be unable to read, write, or understand the English language. He was described as ignorant, uneducated, a child of nature, old and infirm, and without capacity to transact business. His land was estimated to be worth $100,000. The court would rule on April 5, 1915, with the exception of the original lease, that the subsequent leases were inequitable and unconscionable and a fraud upon the allottee. Additionally, having no power to convey his estate in the land, the allottee could not sell that part of the land that consisted of rents and royalties.[656] Effectively, the high court ruled that overlapping leases were illegal and that assignment of future royalties by Indians was illegal.

Charley Quapaw's royalties were returned by Noble and John Cooper's estate. However, such was the influence of Vern Thompson that he persuaded the chief to accept only five dollars for all monies due him from the Thompsons. The Secretary of the Interior sough to enforce full restitution, but Vern Thompson was the Ottawa County Judge at the time, and the U.S. Attorney of the district advised against legal action.[657]

Following his graduation from the Department of Law at the University of Michigan, Thompson relocated in Miami, Indian Territory, in 1902. A county attorney just after statehood in 1907, he would become a county judge between 1912 and 1916.[658] Thompson would later represent mining companies and eventually represent the Quapaw Tribal Council. Much of Thompson's legal practice in later years involved Quapaw related cases.[659] Thompson's professional involvement with the Quapaws would span a period of fifty years and culminate with his representation of the tribe in the *Quapaw Indians vs. United States*.[660] In November 1947, Thompson filed a petition before the Indian Claims Commission requesting a judgment of over $54 million for lands relinquished by the tribe in 1818 and 1824. Previously, the Indian Commission had approved Thompson as counsel for the Quapaws in the adjudication of their claims with the United States Government. Thompson's appointment as counsel had been supported if not prearranged by Quapaw Chief Victor Griffin, but the selection had not been approved in the tribe's open council. The Commissioner of Indian Affairs ignored protests of Thompson's appointment and approved a contingent fee of 10% plus expenses. The claims commission would eventually rule on May 7, 1954, and award the Quapaws over $927,000 from which $107,643 was paid to Thompson for his work and expenses.[661]

Thompson wrote a booklet in 1937 titled *Brief History of the Quapaw Tribe of Indians*.[662] Certainly, Thompson's vision of the Quapaw history and the Tri-State Mining District was different from the more recent and scholarly books by such historians as W. David Baird and Arrell M. Gibson. But, given Thompson's instances of less than fair dealings with the Quapaws as evidenced by the Charley Quapaw royalty affair, his view of the Quapaws' financial acumen and financial status during the boom years of the mining industry does not correspond with the reality of the times. In his booklet he states, "...the fortunate Quapaws having large incomes from the mines, especially members of the older generation were as frugal and wise in their spending as the average white man who had suddenly, and without previous training in care and expenditure of money, come into possession of large incomes."[663] Contrary to Thompson's attestation of the wealthy Quapaws' frugality, another of his stories more correctly paints the typical picture of their spending habits. Thompson tells the story of an old poorly dressed but wealthy Quapaw, with his hair in traditional braids, who had stood around an automobile agency showroom for several hours. Near closing time when the owner noticed the old man, he jokingly asked what car he was going to buy. The old man said, "You all sell George Redeagle a car?" The owner of the agency stated that he had. The old man said, "How many cylinder?" Told that it was a twelve-cylinder Cadillac, the old man replied, "Huh! Him a cheap Indian. Me want a twenty-four cylinder car."[664]

The truth is that there was a rapid dissipation of the fortunes of the Quapaws during the mid 1920s into the 1930s, and the extent of this dissipation is evident when the numbers are examined. Approximately $14.7 million royalty income was received by the Quapaws on their restricted allotments between 1923 and 1943. The bulk of this income was derived in the boom years before 1930. Annual royalties had dropped to slightly more that $83,000 in 1933 and had risen to only $970,000 in 1943 during the peak lead-zinc demand years of World War II. The wealthy Quapaws spent nearly two-thirds of the $8.5 million income derived from royalties during the 1920s. In 1925, royalties were paid to forty-five Quapaws, and the number receiving payments doubled to ninety by 1935. However, only ten families received the majority of the royalties. Concentration of the Quapaws' wealth is evident when one considers that 96% of their royalty income in 1926 went to these ten families. These families included some of the most flagrant spendthrifts of the times. By 1963, only seven Quapaws had balances at the Indian Agency exceeding $45,000 with only two of those exceeding $500,000. This dissipation was aided and abetted by a local government agent, Indian Superintendent J. L. Suffecool, during his tenure from 1924 to 1929. He wrote, "...that there is nothing to be gained and everything to lose by not permitting them to have that which they desire."[665]

David Baird, the pre-eminent historian of the Quapaws, identified the causes of the dissipation of the fortunes generated by royalty income. Shortsightedness and, in many cases, the outright failure of government officials to adequately protect the Quapaws, particularly Superintendent Suffecool during the boom years of the 1920s, were major causes. But Baird states, "...so too were the Downstream people. With a history of abject poverty and a culture that had little concept of saving, or 'deferred gratification,' they valued money only for its ability to satisfy immediate desires." The effects of the fantastic and sudden wealth had a dramatic and permanent effect on their culture. The family and clan system was destroyed. Divorce (some as many as six or seven times), bigamous marriages, and infidelity were the norm among the younger second-generation of wealthy Quapaws. There was an increasing sense of individualism—shunning of common goals. As Baird so correctly put it, "Hence for the Downstream people wealth undermined both family and community, eliminating thereby most of the obstacles that prevented a more complete accommodation to white society."[666]

Following approval by Congress on March 3, 1909, the Bureau

of Indian Affairs issued regulations outlining a process whereby a determination would be made as to the removal of restrictions on Quapaw lands.[667] An investigation was made to determine if the restrictions from sale were to be removed unconditionally and, subsequently, become subject to property taxes. If a Quapaw's allotment was classified as being subject to a conditional removal from restriction, the Secretary of the Interior would dictate sale terms and use of proceeds. Results of the investigation determined that sixty-nine allottees were competent and qualified to have restrictions removed. Another fifty-five were not qualified but were capable of conducting their own business affairs. A third group of twenty-one allottees was considered totally incompetent to conduct any of their business transactions. Abrams feared a loss of ability to lease Quapaw lands if they were ruled to be incompetent and subject to governmental decision making regarding their allotments. He used his influence with the tribal council to get them to oppose the BIA regulations. But he and the council reversed their opposition after realizing that gaining competency status would subject them to paying property taxes. Joining Abrams, Chief Clabber and other Quapaws denounced the BIA's efforts to create a means to remove restrictions "...as unrepresentative of the will of a majority of the Downstream people." However, only two years earlier the chief had derided the government for its position of considering the Quapaw incapable of handling their affairs and "as fit subjects for a Lunatic Asylum."[668]

The issue of taxation of Indian lands would be determined by a U.S. Supreme Court ruling in May 1912. In this case, eight thousand members of the Choctaw and Chickasaw tribes became the plaintiffs against the State of Oklahoma when the state attempted to tax each of their 320-acre allotments prior to the expiration of a tax-free period. The court's ruling favored the plaintiffs and provided that Indian lands would not be placed on tax rolls if restrictions were removed during a period of inalienability or a finding of an Indian's competency.[669] However, to take advantage of this tax-free status, the original allottee or his heirs must retain the land.

The original twenty-five year period of inalienability of Quapaw lands was scheduled to expire in October 1921. Concerned about those Quapaws deemed incompetent, the government formed another competency commission which completed its work in 1921. Of the 336 enrolled members, thirty-one original allottees and thirty-seven allottee heirs were found to be incompetent to conduct their own business affairs, including Chief Clabber. As a result, the commission recommended that approximately one-third (17,225 acres) of the original 1893 allotment continue to be restricted. The Quapaws supported such restrictions and some (including Abrams and Clabber) had lobbied since 1912 to have the restrictions extended. Legislation was approved by both the House and Senate on March 3, 1921, that extended the restrictions until 1946.[670] The act included a provision that the Secretary of the Interior could in whole or in part, with or without an application from the Indian, remove the restrictions "after he has found such Indian owner to be as competent as the average white man to conduct his own business affairs with benefit to himself..."[671] The majority of tribal members were not affected by the extended restrictions, but most of those that were restricted were more than half-blooded.[672]

Only ten years had elapsed since the 1921 extension had been approved when Quapaw Chief Victor Griffin requested Congress extend restrictions, not scheduled to expire for another fifteen years, prohibiting alienation of the restricted Quapaw allotments. But it would be 1939 before Congress would pass legislation and be signed by President Roosevelt to extend the BIA governance and restrictions until 1971.[673]

Louis Angel, known as Tallchief, was the last of the Quapaw hereditary chiefs. Following Tallchief's death in 1918, the tribe elected its chiefs from its members. Peter Clabber was elected following Tallchief's death and served from 1918 to 1926. John Quapaw served only two years from 1927 to 1928, and was followed by Victor Griffin who served twenty-nine years from 1929 to 1958. In 1956, the Quapaw Business Council headed by a chairman began handling all tribal activities, administration, and financial matters. Others in the seven-member council are a vice-chairman, secretary-treasurer, and four council members.[674]

Robert Whitebird was a boy of eight when he was listed in the 1921 extension of restrictions on certain Quapaws.[675] Following a reorganization of the tribe in 1956, Whitebird became the first chairman of the Quapaw business committee. He and his colleagues campaigned for continued government supervision and restrictions. Although Whitebird retired in 1968, his lobbying efforts would pay-off in 1970 when Congress approved a third twenty-five-year extension of the government's role as the tribe's federal guardian on almost 12,500 acres. The extension expired in 1996, one hundred years after the original allotments were granted and the imposition of the first twenty-five year restriction.[676] Born in 1913, Whitebird was the last full-blood Quapaw at the time of his death in May 2005.[677]

Over the years, the Indian office attempted to conserve Quapaw Indian wealth through administration of personal finances and leasing regulations. The Indian office also opposed all taxes on the Quapaws—federal, state, and local. Litigation was another weapon the office used to protect the Quapaw from those who had cheated or defrauded them. Consistently supported by the Quapaws, this protection fostered a continued dependence on the government but was in conflict with the growing individualism resulting from great wealth and assimilation into the larger society, an ever-present and over-riding concept promoted for decades by the United States Government.[678]

With a swiftly expanding global population and extensive migration of people from all nations around the world, ethnically distinct populations are being biologically assimilated with people of other ethnic origins at a speed never before known. The Indian world is a part of this blending process, and therefore the Indian blood quantum of the descendants of the various tribes becomes smaller with each succeeding generation. This has happened to the Quapaws.

One may call himself an Irishman even if his Irish ancestors only connect several generations back and represent only a small fraction of his heritage. We may smile at our proud Irish heritage-claiming friend. However, unlike other ethnic groups, American Indians with official citizenship in a tribal group legally recognized by the United States Government have as individuals a recognized ethnic status. The Indian may have no more Indian blood than our Irishman has Irish blood, and in fact the Indian may have more Irish blood than Indian blood. Yet the Indian is legally recognized as part of a tribe, but the Irishman only has some interesting stories about his Irish connection to tell around the family dinner table.

In 1890, the Quapaw roll listed 193 persons, including many non-Quapaws adopted into the tribe. The 1910 census of the Indian population in the United States shows a remarkably high percentage of Quapaws with mixed white and Indian blood (including adopted Indians from other tribes). The Quapaw population in 1910 was 231, only thirty-eight more than the 1890 population census, and only ten Quapaws lived outside of Oklahoma in 1910. Of the 231 Quapaws shown in the 1910 census, 103 were half-white or more than half-white. In other words, 45% of the 1910 Quapaws were half-white or more. The relatively large numbers of Indians with a high percentage of white blood was not unusual for many if not most tribes listed in the 1910 census.[679]

Today, there are approximately 3,100 enrolled Quapaws who have been able to claim kinship with at least one of those Quapaws listed on the 1890 tribal rolls. Because of shrinking tribal rolls, the tribe does not require a minimum level of Quapaw blood quantum in order to be an enrolled member of the tribe.[680] The Quapaws have survived as a recognized people group. But, like our Irishman and in spite of the official tribal membership card carried in their wallets, the connections of many of today's Quapaws with those who arrived in 1830s Oklahoma are tenuous at best.

EPILOGUE—JOURNEY'S END

"But her end is bitter as wormwood, sharp as a twoedged sword. Her feet go down to death; her steps take hold on hell."[681]

Our journey began with a Saturday morning bicycle ride through the pitted prairie on dust-choked roads surrounded by the chat hills of Picher and Cardin. The steady crunch of loose gravel beneath the bicycle's wheels was a reminder that this was not only a journey of discovery of the present but also a call to the past. The rubber tires were merely riding on the rim of the present, as though we were at the rim or edge of the universe traveling away from the center of its creation light years in the past.

As we face the vast, black unknown of the future, how can we know, understand, and tell the story of this present moment without looking over our shoulder at the past? We cannot, and we could not have told this story without examining the history of the land, people, and events that culminated with this horribly fascinating landscape called Tar Creek.

As Claude G. Bowers has stated, "History is the torch that is meant to illuminate the past to guard us against the repetition of our mistakes of other days. We cannot join in the rewriting of history to make it conform to our comfort and convenience."[682] Our story may not be a blazing torch but hopefully will be a small glowing ember that will enlighten some hidden pockets of our past and the nuggets of truth lying therein. Bowers' hope that knowledge of history will inoculate us from the repetition of our mistakes may be true. But, in our modern world, the vaccine of history has been relegated to benign neglect at best or perverted at worst by our educational system in the United States. The intelligentsia, media moguls, and the social and political elite have a long and consistent record of manipulating history to accomplish their own agendas, especially in today's raging cultural and political wars. And for centuries history has been perverted or buried by the world's despots and dictators, and when history is twisted or perverted, it is not truth and therefore fails to offer the guidance sought by Bowers.

Truth is absolute and unchanging or it is not the truth. Truth stands the tests of time. Truth may be ugly, beautiful, terrible, or wonderful. Truth is what it is. As humans, we must separate the kernels of truth from the chaff of fictions, fabrications, and follies disguised as truth. But truth is not enough. Truth is merely the grist for the refining mill of our minds and spirits. We must synthesize the truth and act on our knowledge and insights. The fine flour resulting from this milling process will expand our understanding of the important issues of life and nourish our souls.

Unfortunately, much of history and the truths it presents are relegated to the dusty shelves of libraries, ignored by all except the dedicated historian. King Solomon's charge to "Buy the truth, sell it not..."[683] seems out of place in the modern age of computers, instant communication, and electronic media presenting sound bites of inane trivia. So why bother with history? The bother is that truth will survive and act as a guiding light for those who seek it. Like the kernels of grain found after being buried in Egyptian tombs for millennia, truth will blossom and grow as it is exposed to the light of our knowledge of history and is watered and nourished in the fertile soil of our minds and spirits.

The writer has made certain observations that the reader may examine and accept or discard. Whether or not the microcosms of history presented in this book provide a sound basis for discovering certain truths and insights is left to the reader's interpretation. There is always danger in ascribing certain truths based on limited evidence. However, an accumulation of evidence eventually will provide a solid basis that will reveal those truths, and these observations are small deposits to that body of evidence. The first observations deal with the importance of marriage, family, and community and the role that these entities play in achieving cooperation and unity in a civil society. The second examines the American experience with the pioneering spirit, the rugged individualist, the self-made man, and the consequences thereof. The third insight involves government and the problem of balancing the pressures of media, interest groups, and the public at large in accomplishing the business of governing. The subjects will receive little more than cursory attention. Volumes could be and have been written on each of the topics. But again, these topics are merely observations and questions posed as grist for the reader's mind. Let us begin.

We followed the account of a people as they escaped their antagonists by moving down the Ohio and Mississippi Rivers. After several centuries, the Quapaw would eventually inhabit a tiny reservation in far northeastern Oklahoma. Even as their survival was in doubt at the close of the nineteenth century, events were occurring that offered hope as well as disaster. The history of the Quapaw appears to be replete with missed opportunities, wrong decisions on the part the people and its leaders, and a generally slow disintegration of tribal cohesion over the centuries.

With the flight down the Ohio River to escape the murderous Iroquois, the Quapaw, Osage, Kansa, Ponca, and Omaha arrived at the Mississippi in the late 1600s. This may have been one of the most important turning points in the Quapaw history. Their Dhegiha Sioux cousins proceeded upstream, but the Quapaw, lost in a fog bank, drifted downstream and became known as the Downstream people. It is not known whether this casualness about separation from their kinsmen was a part of their nature cultivated over millennia or was of recent origin. However, just as their canoes drifted along with the currents of the Mississippi, their cultural drift foreshadowed a similar lack of cohesion as a community in numerous instances in the centuries to come. Time after time we see the social disintegration of the Quapaws as they took the road of expediency and self-interest as opposed to the interests of family, tribe, and community.

The Quapaws settled near the confluence of the Arkansas and Mississippi Rivers. The ease with which the Quapaws separated themselves from their Siouian cousins was also evident in their marriage relationships. As previously discussed a Quapaw husband could walk away from marriage if he deemed his wife to be loathsome to his family. Likewise, a wife's father could retrieve his daughter if he perceived the husband had mistreated her.

Not only were the Quapaw quick to end relationships, they were quick to establish relationships, many times to their misfortune. In 1673, just a few years after their arrival at the mouth of the Arkansas, the tribe had their first encounter with the white man. The Quapaw quickly embraced trade with the French and Quapaw women would eventually inter-marry with the French. In fact, such was the inter-marrying of the Quapaws and French, a visitor to the Arkansas Post in the very early 1800s would have been immediately struck by the considerable extent to which the residents were of mixed blood and spoke a crude mixture of French and Quapaw.[684]

The lack of tribal cohesion and unity among the Quapaws was especially evident in the nineteenth century. Following the Louisiana Purchase, westward expansion of the United States would put pressure on the Quapaws to vacate their land to white settlers and move westward. In early 1826, the Quapaws were forced from their Arkansas home to distant Caddo lands in Texas just across the Red River. By September of that year, starvation and hardship caused one-fourth of the tribe to return to Arkansas. The remainder would eventually join the Cherokees in Texas or reside with the Choctaws in Indian Territory. In the 1830s, the Quapaws that returned to Arkansas would be moved to a reservation in the far northeastern corner of Indian Territory. But, by 1876, half of those Quapaws residing on the reservation had moved westward to reside with the Osage on their reservation.

The importance of the family unit and community cannot be over emphasized. William Bennett brilliantly and succinctly sums up this importance.

> Marriage and family are cultural universals. Everywhere, throughout history, they have been viewed as the standard to which most humans should aspire. This is not happenstance; it is, rather, a natural response to basic human needs—basic to individuals, and basic to society.[685]

We frustrate the fulfillment of these needs and imperil our society when family, and consequentially community, is devalued by the widespread loss of preeminence and respect or outright dissolution. Marriage and family form the essential elements in the intricate landscape of a community and nation. Again quoting Bennett, this intricate network brings "…certain built-in expectations, reciprocal obligations, and formal responsibilities."[686] History books are packed with many examples of the power of unity and solidarity of families and communities of people. Where this is lacking, writers have chronicled the lives of people groups such as the Quapaw that are only a shadow of their former selves.

The Quapaws' historical lack of a cohesive society is a microcosm of the conditions we find in the western world today. In many ways, the ills of modern society can be traced to the disintegration of the family unit and the unifying force it once brought to community and society as a whole.

One of the great social disasters of the last forty years in the United States and other western countries is the disintegration of the family unit. Until his death in 2003, former United States Senator from New York Daniel Patrick Moynihan was one of the most thoughtful, articulate, and respected leaders in the country. As an assistant secretary in the Department of Labor in 1965, his office issued *The Negro Family: The Case for National Action.* That report became known as the Moynihan Report and concluded that African-American urban poverty was due in part to a breakdown of family structure. Although highly criticized at the time, the report's conclusions proved to be prophetic.[687] Upon his retirement in 2000, Moynihan was asked what the biggest change he had seen in his forty years in politics was. Moynihan responded, "The biggest change, in my judgement, is that family structure has come apart all over the North Atlantic world." In support of his statement, Moynihan pointed to the illegitimacy ratio in the United States, Canada, Britain, and France averaged between 5 and 6% in 1960. By 2000, the ratio averaged between 34 and 35%.[688] The great increase in single-parent households coupled with the economic and social pressures of modern life have created a culture in crisis. Drug addiction, alcoholism, domestic violence, child abuse, an increase in the number of abortions, crime, and greater social acceptance of homosexuality are only a few of the consequences of disintegration of the both the immediate family unit as well as the ties between generations of the family unit.

Politicians and social engineers debate solutions for the problem. The weapons of choice are political, economic, and social. Indian tribes including the Quapaw have sought resolution of their problems through political efforts (tribal sovereignty) and economic endeavors (primarily gambling enterprises). Neither addresses the root cause of the social ills of Indian society. In fact, both greatly aggravate family disintegration in the Indian world and as well as the rest of the nation.

Tribal sovereignty has become a political tool used to further the interests of tribal leaders and their supporters. While some benefits flow to tribal members, the benefits to a large majority of tribal members are minimal. Even those few tribal leaders who have the interests of the tribe at heart must realize that the short-term benefits are far out-weighed by the long-term costs. The social and economic damage to the country as a whole cannot be overestimated. Whatever the economic gain and political power achieved by a tribe, the creation of a nation within a nation is divisive to the extreme, and divisions will only widen as the recognition of tribal sovereignty and its consequences spread throughout all levels of economic and social interaction. Special interests and privileges and the inequalities created thereby (be it tribal sovereignty or affirmative action) are destabilizing elements in any civil society that will lead to a greater social and political fracturing of that society.

Tribal sovereignty has also made possible the rapid spread of gambling in the United States. Twenty years ago legalized gambling was permitted in only two states. Now, only two states remain free from some sort of commercialized gambling.[689] Added to this is the explosive growth of gambling over the Internet. The economic and social costs of gambling are staggering. Not even the staunchest of the industry's defenders can deny the problems caused. In 2004, 367 tribal gaming operations generated $19.4 billion in revenues[690] compared to $10.6 billion for Nevada casinos.[691] The growth of the Indian gaming industry is being outpaced only by the social and economic problems created by it for individuals, families, communities, and the nation. It is estimated that 2.5 million adults in the United States are pathological gamblers at a cost of $5 billion per year with lifetime costs of $40 billion for reduced productivity, additional social services, and creditor losses. Another three million are considered problem gamblers.[692] The greater costs lie in the area of family and include domestic violence, child abuse and neglect, crime, suicide, and depression. Again, these factors cause disintegration of the family and result in other societal problems as previously enumerated.

A second area for consideration is an examination of the American experience of what is variously called the pioneer spirit, the rugged individualist, or the self-made man mentality that has been both a great contributor to the development and prosperity of the United States as well as a cause of harm in many ways to individuals and society as a whole.

Our first hundred years proved us a nation on the move, always looking for new opportunities. We were a new nation, a land of opportunity. We had thrown off Britain's yolk. More importantly, we had thrown off the mindset that what station or social class a man was born to would be his lot in life. In the United States, a man could become what he wanted to be if he worked hard enough and mixed in a little luck too. The western horizon beckoned. There was land to be had and fortunes to be made. A prospector one day, a millionaire the next. Anything was possible. Thanks to these pioneering, entrepreneurial men and women, the United States is now the most prosperous and powerful nation on earth. And in addition to power and prosperity, we also became a good and charitable people.

So it was in the lead and zinc-laden hills of the eastern Ozarks and gently rolling prairie to the west near the intersection of Missouri, Kansas, and Oklahoma. The pioneer spirit was alive and well as the nineteenth century came to a close.

The Tri-State Mining District by its very nature was ideal for the culturally prepared lone prospector or small operator. All it took was a minimum of equipment and a strong back. A man need depend on no one except possibly a like-minded partner to operate the windlass as he descended into the shaft. Even with the rise of the large mining companies following the Picher strike, many worked in the mines just long enough to get enough money to strike out on their own. One of the great hindrances to unionization in the district was the prevalence of this independent, pioneer spirit. Many of these poor miners had once worked beside friends who had become wealthy.

Our first observation or insight reflected on the importance of unity of family, community, and national unity—all essentials in weaving the fabric of a civil society. Our second observation that individualism has been and continues to be important in the development of this nation appears to be in conflict if not contradictory with our first supposition. Unity versus individualism—how do we reconcile this apparent contradiction? This conflict was recognized early on in our nation's history. Alexis de Tocqueville wrote the classic *Democracy in America,* one of the most highly influential books on political thought since publication of the final volume in 1840. In his book, Tocqueville argues that unlike the aristocrats, despots, and dictators of Europe, American democracy was not dependant on a chain of obligation from peasant to king. Democracy broke the chain.[693] The Americans' brand of democracy placed all men on an equal basis, a level playing field in today's vernacular, where self-reliance and independence were the order of the day and bred a uniquely American individualism. The past and future counted for little compared to the here and now for these individualists.

However, Tocqueville recognized the dark side of individualism in his writings.

> Individualism is a mature and calm feeling, which disposes each member of the community to sever himself from the mass of his fellow creatures; and to draw apart with his family and friends;

so that, after he has thus formed a little circle of his own, he will-
ingly leaves society at large to itself.[694]

For democracy to work and survive in the long run, something
must be present to rein in this rush to unrestrained self-interests
and disengagement from society. Indeed, America has succumbed
in times past to destructive self-interests by allowing a slave-holding
portion of society to continue four score and seven years after the
birth of the nation. America tolerated the existence of the sweat-
shops where women and children toiled long hours during the late
nineteenth and early twentieth centuries. We remained immobi-
lized during the beginning years of the Great Depression for lack
of a historical template to fix the tragedy of human deprivation that
self-reliance and individualism couldn't address. But in spite of the
failures of our democracy, can the masses of any society desire the
alternative of rule by aristocracy, despotism, or dictatorship?

So why hasn't this peculiar form of American individualism and
self-interests toppled the nation from the mountain of democracy and
envy of the world into the pit of bondage of non-democratic forms of
rule? Tocqueville believed that the dangers of unrestrained individualism are held in check by enlightened self-interests. Tocqueville stated
it as follows:

> When the members of a community are forced to attend to
> public affairs, they are necessarily drawn from the circle of their
> interests, and snatched at times from self-observation. As soon
> as a man begins to treat of public affairs in public, he begins to
> perceive that he is not so independent of his fellow-men as he had
> first imagined, and that, in order to obtain their support, he must
> often lend them his cooperation.[695]

In other words, men recognize that one's self-interests are
advanced when actions are taken in cooperation with others. In
essence, one prospers when one cooperates. Tocqueville found that
this cooperation occurred through the formation of a multitude of
political, civic, and social associations and organizations.

Is this enlightened self-interest significantly declining or disappearing from the United States? If so, why is this occurring? We have discussed Tocqueville's position that enlightened self-interest results from the citizenry getting involved in local affairs that brings citizens out of their isolation and reminds them that they are an integral part of the society in which they live. In other words, that citizenry recognizes that they are responsible for being a part of the solution, and it is in their best interest to accept this responsibility. Again, one need only go back to 1960s to see the beginnings of the move away from enlightened self-interest to a greater reliance on federal and state governments for the solution to all sorts of social and economic ills. Some may argue that the first wave of this move occurred in the 1930s with the massive intervention of the federal and state governments to alleviate the suffering nation from the effects caused by the Great Depression. However, it was enlightened self-interest that resulted in vast governmental programs to feed the hungry and jumpstart a prostrate economy. But, just as a wonder drug may relieve pain and promote healing, there is a tendency to overmedicate that is followed by a growing addiction to the very thing that was so valuable in restoring one's health in the beginning. Unlike the enlightened self-interest of the 1930s, this addiction to pervasive governmental involvement in the affairs of its citizens began manifesting itself in the 1960s with a multitude of programs labeled as The Great Society. Subsequently, the federal and state governments were perceived to be responsible for and therefore must attempt to perform the tasks that the multitude of associations (local government, societies, and organizations) had once endeavored to attend. As a consequence, individuals tended to lose that sense of their ability to achieve common goals.

Even in the 1830s, Tocqueville foresaw the growing complexity of a modern society. To deal with this complexity, citizens must band together in various forms of associations (business and civic organizations, societies, and local government) or increasingly rely on federal and state governments. Tocqueville wrote,

It is easy to foresee that the time is drawing near when man will be less and less able to produce, of himself alone, the commonest necessaries of life. The task of governing power will therefore perpetually increase, and its very efforts will extend it every day. The more it stands in the place of associations, the more will individuals, losing the notion of combining together, require its assistance: these are causes and effects which increasingly engender each other.[696]

To restate Tocqueville's proposition, the more a government does for its citizens, the more those citizens will need assistance of that government.

With regard to the principle of cooperation and unity of purpose through enlightened self-interest, let's return to the history of the Quapaw. In their early contacts and relations with the Europeans, both the Quapaws and Europeans unknowingly practiced the concept of enlightened self-interest later defined by Tocqueville. As Morris Arnold so effectively describes, the 125 years of symbiosis between the Quapaws and the French and Spanish was achieved as they made accommodations and compromises with respect to their cultures. Just as the Quapaws were adapting, assimilating, and changing as they absorbed European influences, so too were the French and Spanish doing likewise with regard to the Indian cultures surrounding them. Small in numbers and a half a world away from the imperial courts of Europe, the French and Spanish leaders in the New World had become Indianized, in effect Indian chiefs and not European rulers.[697] In particular, the French had come to depend on their symbiotic relationship with the Quapaws by the middle of the 1700s.[698] Given the Quapaws' strategic location on the Mississippi and Arkansas Rivers and the generally high regard with which the Europeans held them, there is perhaps no other tribe that was as successful in meeting and adapting to the challenges brought by the Europeans presence prior to 1800. For various reasons discussed in this book, the Quapaws appear to have abandoned the principle of cooperation and unity of purpose through enlightened self-interest in the nineteenth century. In addition to the fragmentation and dispersion of the tribe during the nineteenth century, their handling of the enormous wealth discovered under their northeastern Oklahoma reservation is a further example of the abandonment of this principle.

It has been noted that the great oil discoveries and discoveries of enormous lead and zinc deposits, both in northeastern Oklahoma, occurred almost simultaneously on Indian reservations separated only by a few miles. The Osage sat on huge oil deposits while the Quapaw resided on vast deposits of lead and zinc. Today, the Osage are one of the richest tribes in the nation while many Quapaw live in poverty. The difference resulted from the manner in which the mineral wealth was owned and distributed. All Osage mineral deposits were held in common. When oil was discovered on the land of one Osage tribal member, all prospered. To the contrary, the Quapaw owned their minerals individually, and while a few obtained extreme wealth if they were fortunate to have lead and zinc deposits on their allotment, other less-fortunate Quapaw neighbors would starve in extreme poverty. An additional benefit of the Osage tribe's enlightened self-interest in pooling their mineral wealth was the protection of the corpus of that wealth even though an individual may lose through fraud, swindle, and treachery. Not so for the hapless Quapaw millionaire. When his wealth was gone, there was nothing with which to replace it at the next annual disbursement.

This principle of cooperation and unity of purpose through enlightened self-interest was also lost on the mining community— on both the operators and miners alike. The dismal record of unions in organizing the Tri-State was due in large measure to a particularly strong measure of individualism bred into the inhabitants of the lawless prairie and isolated hills of the Ozarks, which had not been tempered by realization of the value of unity and cooperation. Likewise, many of the operators, owners, supervisors, and foremen were born and raised in the area and failed to see the advantages of cooperation and unity with their employees, again the result of the absence of an enlightened self-interest.

Our third consideration or insight involves government and the problems of balancing the pressures of media, interest groups, and the public at large in accomplishing the business of governing. We have talked of an increasing reliance on federal and state governments to address the staggering problems faced by an increasingly complex society. We have learned that such dependence causes individuals to rely less on local governments and other associations in addressing the needs and problems of society. There will soon be a third generation of Americans that has grown up in a society looking to the federal and state governments almost exclusively for answers to life's problems. Whether it is hurricane damage relief, crime in the streets, personal health, or a myriad of other issues, a large portion of the population believes that the federal government is supposed to supply the answers and funding for addressing an ever-expanding list of needs and problems. Such a mindset—promoted for decades by the major media, vested institutional interests, and special interest groups—tends to become the accepted gospel…the way life must be. There is no collective memory of a better way.

That said, the Tar Creek Superfund site cleanup is not something that local organizations are capable of handling. However, the local and state governments and other organizations, aided by the media, have also adopted the mindset that Washington has all of the answers and must accede to their demands. This type of adversarial role assumed by these entities has delayed effective action on this monumental problem. One recent example of these problems involves media coverage of the Tar Creek cleanup effort. The *Tulsa World* appears to have had a predominately political agenda for the better part of three years beginning in mid 2003 in reporting and editorializing about Tar Creek. As the only major daily in eastern Oklahoma, the publishers and editors had a free hand in developing a campaign to direct events of government to suit their political agenda. This is not to say that government inaction and waste should not be vigorously challenged. But the actions of this newspaper with regard to Tar Creek are only a microcosm of the overwhelming and damaging power of major media in this country. The power and agendas of such major media, vested institutional interests, and special interest groups tend to undermine and frustrate thought-ful and considered actions of government. Nor is media dominance and misuse of recent origin. The *Arkansas Gazette's* campaign in the 1820s to have the Quapaws removed from Arkansas is another excel-lent example of journalism gone awry.

How did we arrive at this state of affairs? Once again we must return to Tocqueville. Of the newspapers he observed in the 1830s, Tocqueville wrote,

> To suppose that they (newspapers) only serve to protect free-dom would be to diminish their importance: they maintain civi-lization. I shall not deny that in democratic countries newspapers frequently lead the citizens to launch together in very ill-digested schemes...The evil which they produce is therefore much less than that which they cure.[699]

During his tour of America, Tocqueville found an enormous number of newspapers. It was Tocqueville's contention that the num-ber of newspapers was not the result of great political freedom of the country or the freedoms enjoyed by the press. Rather, the multitudes of papers were the direct result of the considerable subdivision of administrative power in the country. In other words, the greater the decentralization of government is, the greater the number of news-papers that are needed to report the affairs of the day. Conversely, the more centralized that a government becomes, the fewer number of newspapers that are needed.[700] Others challenged Tocqueville's explanation that the abundance of American newspapers was a result of the decentralization of American government that required local papers to carry news of local interests. Those detractors pointed to the relatively small amount of local news in comparison to national or world news. However, volume of local news does not refute Toc-queville's observations. Even if Tocqueville's detractors were correct, it was the democratic nature of American government that caused the populace to hunger for news so that they could be better pre-pared to participate in the affairs of the nation.[701]

The prevalence of newspapers and significant readership of the era are striking. By 1835, the population of the United States had grown to fifteen million, almost four times the 3.9 million in 1790. However, during the same period, the number of newspapers increased almost twelve-fold from 106 to 1,258. In the 1780s, there were just under twenty subscriptions to newspapers per one hundred households compared with fifty subscriptions per one hundred households in the 1820s. Although there were many multiple subscription households, the general spread and saturation of newspaper readership was remarkable, especially for a largely agrarian society.[702]

In an age where newspapers are declining in number and those that remain are steadily losing circulation, do Tocqueville's suppositions remain true? First, we must recognize the mammoth changes in how Americans receive their information. Beginning in the early 1900s, journalists were much more likely to become the independent interpreters of the political scene and to put forth their own analysis. In the 1920s, Walter Lippman wrote of a crisis in journalism in his books on the American media. Following the manipulations of the press and false news reports during World War I, Lippman coined the phrase "the manufacture of opinion" to describe the detestable elements he saw in American journalism. The 1920s saw the rise of syndicated journalists and radio commentators commanding large audiences and wielding considerable influence. Lippman called for journalists to become professionals, to deliver trustworthy and relevant news, to be objective, to refuse to withhold news or put moral uplift or any other cause ahead of veracity, and to be held accountable. Beginning in the 1930s, many would attempt to debunk Lippman's conclusions and prescriptions.[703] But, such criticism could not minimize the truth of the media's power to shift culture and practices over the longer term and to cause "the narrowing of ideological diversity."[704] Fortunately, this narrowing of ideological diversity has been countered somewhat by the growth of talk radio and the Internet in the 1990s.

The domination of newspapers began eroding with the advent of radio in the 1920s. Television accelerated this decline beginning in the 1950s. The advent of the Internet continues to hasten the decline in the numbers, circulation, and importance of print periodicals. As a result, many newspapers have consolidated or merged into extensive business enterprises that control both print and electronic media. The drift toward centralization of American government since the early part of the twentieth century has paralleled the consolidation of the American media during the same period.

Tocqueville was not able to foresee the dramatic changes in print media, the invention and development of various forms of electronic media, and the extent of the centralization of American government. The new journalistic landscape effectively dominates and centralizes the delivery of the news and therefore forms the attitudes and opinions of millions of Americans by the few. The print and electronic media, both historic defenders of the free press and freedom of speech, are effectively accomplishing the suppression of those ideals not by forbidding free speech and a free press but by muffling other viewpoints by the sheer scale and extent of their reach. The small town newspaper may trumpet opinions that contradict the positions of the *New York Times, Washington Post,* and the major television networks. But for every one of those small town citizens that reads the local paper, many more will listen to the evening news presented by their favorite national news anchor. Also, it is naive in the extreme to talk about fairness in reporting, journalistic integrity, and the like. In spite of Lippman's cautions, every reporter, every editor, and every media owner has an opinion, and those opinions bleed through their journalistic products to the consuming public regardless of how carefully each performs his job. Even that care has been lacking for many years as journalists (from owners to the lowliest reporters) have blatantly stated agendas to educate the public to view issues in a certain way, to champion a cause, or promote a certain political agenda. As one writer expressed his concern,

> At its origins, liberal democracy cherished the press as a public guardian, little anticipating its metamorphosis into a powerful industry with its own imperatives. In the twentieth century, particularly in the aftermath of World War I and other developments that raised concerns about manipulation of public opinion, some

critics began to ask how to reconcile democratic ideals with the media's power and limitations.[705]

Our past protection from this demagoguery has been the great diversity of opinion that has been reflected by thousands of media sources. But with the present state of consolidation resulting in an all-powerful and all-knowing media, those agendas that should appropriately be relegated to the editorial page or evening news commentary are standard journalistic reporting fare with most of the large news dispensing organizations.

There is an ironic and interesting parallel between the modern media behemoths and the development of telecommunications in Russia following the 1917 revolution. Until its collapse in 1991 and irrespective of the differing levels of affluence, the Soviet Union and its satellites had substantially fewer telephones than North America and its European neighbors. The new Soviet leaders of 1917 consciously invested in loudspeakers, not a telephone system, and subsequently broadcast technologies, another form of loudspeaker. The totalitarian rulers were more interested in disseminating information to its citizens as well as limiting communications between citizens.[706] In spite of the protestations of today's media giants as they wrap themselves in the self-righteous tunic of fairness, free speech, and a free press, a peek beneath this tunic reveals the harsh blare of the bullhorn held by a mailed fist.

There has been a dramatic shift in American journalism since Tocqueville wrote, "In the United States each separate journal exercises but little authority, but the power of the periodical press is only second to that of the people."[707] In many respects, it would appear that in today's world the power of the media has now surpassed the power of the people.

So, what is being said? Point one emphasized the importance of family and community that fosters a cooperative and united civil society. Point two champions American individualism. These two seemingly incompatible concepts have been tempered and welded together by a brand of enlightened self-interests promoted and enabled by the American passion for forming cooperative associations of all kinds to address the needs of society and simultaneously serve individuals' self-interests through a non-centralized style of government. Point three addresses the loss of a diversity of opinion and thought through the consolidation of American media into an 800-pound gorilla sitting at our breakfast and dinner table who dispenses a narrow set of opinions and drowns out competing thoughts and ideas. Such loss of diversity of opinion inhibits and complicates the business of governing by those charged to do so.

Our national obsession with individual rights and personal freedoms has gained ascendancy over the preservation of our nation's interests. Individual rights and personal freedoms are important. At the same time, concern for our country, its ideals, and place in the world are of little importance to a growing number of Americans. In sharp contrast to previous generations before 1960, the response now is, "What's in it for me?" "What will it cost me?" Or, "It's not convenient." Political expediency and advantage and personal agendas have steam-rolled political and social unity. For a short period following the tragedy of September 11, 2001, the nation regained this political and social unity. It was a fleeting moment in our history when the shock of events caused us to put aside political and personal agendas and recognize that we are one nation, not just a collection of ethnic groups, political parties, and business interests. But such unity built on emotion lasts only for a brief period. The foundation of family and community unity that fostered this national sense of unity and cooperation has eroded over the last forty years and is no longer adequate to rein in the unrestrained pursuit of power and social and economic advantage to the detriment of the nation.

To our great harm, the importance of family and community unity is a concept foreign to many in this country. Beginning in the 1960s, we have become the self-absorbed "me" generation, and as the "me" generation decreases its involvement in those local governments and associations for their enlightened self-interest, the more it must depend on federal and state governments for assistance. This disengagement from the governing process, coupled with ignorance of history and the price paid by previous generations for our rights and freedoms, is aggravated by low voter participation, particularly among the young.

One may become depressed and despair for our nation while reflecting upon the seeming loss of the importance of the family unit and the loss of a united and cooperative civil society. What can turn the tide? One must remember that a relatively small rudder can turn the largest ship. The petite rider through the power of bridle and bit can direct a galloping 1,500-pound horse. But is it possible to change the direction of a whole society from its rush to oblivion? Recall the ideas of James Burke presented in the forward to this book. The small, seemingly innocuous occurrence or act may have gigantic repercussions hundreds of years later and other seemingly monumental events may lead to the ordinary. Two examples have been examined in this book. The seemingly small, insignificant occurrence of separation of the Quapaws from their kinsmen when they drifted downstream had monumental consequences for their future hundreds of years later. The fantastic discovery of lead and zinc under the Oklahoma prairie promised amazing wealth and a bright future for the region and its inhabitants. However, that monumental event has lead to a ravished land populated by a remnant of impoverished people struggling to survive.

In keeping with Burke's thought that small actions may have gigantic repercussions, our small seemingly powerless words and acts that stand in opposition to the rush of a society gone the wrong path may yield extraordinary results. But perhaps Burke leaves too much to chance. Tocqueville's words bring comfort and better expresses our position.

> Thought is not, like physical strength, dependent upon the number of its agents; nor can authors be reckoned like the troops that compose an army. On the contrary, the authority of a principle is often increased by the small number of men by whom it is expressed. The words of one strong-minded man addressed to the passions of a listening assembly have more power than the vociferations of a thousand orators; and if it be allowed to speak freely

in any public place, the consequence is the same as if free speaking was allowed in every village.[708]

Using Tocqueville's template, our nation can restore the importance of and respect for the family unit and re-establish a unified and cooperative civil society. How? By relying on the authority of the principles upon which this nation was founded, relentlessly espoused by a few strong-minded men and women, the course of the nation may be righted just as the course of a giant ship can be changed by its tiny rudder. Tocqueville's words remind us of Isaiah's prophecy 2,700 years ago. "A little one shall become a thousand, and a small one a strong nation: I the Lord will hasten it in his time."[709] Therein lies our hope.

ADDENDUM—THE NORTH AMERICAN INDIAN—
ANCIENT ORIGINS

*"So the Lord scattered them abroad from thence upon the face of all the
earth..."*[710]

There has been and continues to be considerable debate regarding the origins of mankind, the length of time on this planet, and his migrations. Estimates of our sojourn range from 6,000 or 7,000 years ago to hundreds of thousands of years in the past, and the author does not propose in this work to agree with or promote a particular viewpoint. What follows represents a general synthesis of many of the theories proposed by those who have studied the glacial periods and man's migration to the western hemisphere.

The mists of times past have parted very little in the last century. However, paleontologists, archeologists, historians, and ethnologists have proposed various theories and speculations about the pre-history of the western half of the world that are based on notable research and significant archeological discoveries. Their efforts have allowed us a glimpse, however dimly, through these mists into the pre-history of this hemisphere and the trail that led the Quapaw people to north-eastern Oklahoma.

It is a fairly well-accepted premise that man did not originate in this hemisphere but migrated from eastern Asia over the land bridge from Siberia to Alaska.[711] Evidence of human activity in the hemisphere is supported by projectile point finds associated with mammoth and bison kills around Clovis, New Mexico, that date between 11,500 and 11,000 years ago. But beyond this general consensus, there is considerable debate as to when and how these migrations occurred.[712]

During the Pleistocene Ice Age, two ice fields covered the northern portion of the American continent. The mountains of western Canada and surrounding area were covered by Cordilleran ice. The much larger eastern ice field, known as the Laurentide ice, centered on the Hudson Bay area. When glaciation was at its greatest, the two ice sheets joined and formed a single sheet stretching from the Atlantic to the Pacific totaling six million square miles in area, which is larger than the Antarctic ice sheet of today.[713]

Beringia was that broad land bridge between Asia and Alaska that probably existed twice in recent pre-history.[714] Whether over a true land bridge or by boat across a much shallower Bering Strait, radio-carbon dating indicates that the first migrations from Asia occurred between 28,000 and 26,000 years ago.[715] The climate was milder than it had been or would be in several millennia to come, as a final assault of the Ice Age would occur. There is evidence that this final thrust of the ice cap reached its maximum extension southward between 17,000 and 14,500 years ago. Around 14,000 years ago and thereaf-ter, the climate began to warm. By 12,000 years ago, the land bridge had disappeared as rising ocean levels blocked further migration by land.[716] With this last climate change and warming, corridors began to develop in western Canada between the eastern and western ice fields. These corridors stretched from the Yukon border and Northwest Ter-ritories to the southeast through Alberta and to the Great Plains of the United States. A nomadic people followed these corridors south-ward and developed a big-game hunter tradition or culture.[717]

Called Paleo-Indians, this nomadic people hunted mammoths, mastodons, and a bison larger than those hunted by the later Plains Indians. This culture was present at least 12,000 years ago and pos-sibly much earlier. With the demise of the mastodons, mammoths, and other large animals beginning 9,000 to 10,000 years ago, the big-game hunting culture also began its decline but survived in the Plains to as late as 6,000 years ago.[718]

Study of dental traits and language groupings of Native Ameri-cans have suggested three waves of migration to the New World. Native American languages have been classified in three groups—the Eskimo-Aleut in the high Arctic and coastal Alaska, the Na-Dene on the northwest coast and interior Alaska, and the Amerind in the remainder of the continent to the south. It is believed that the

MAN'S MIGRATION TO THE WESTERN HEMISPHERE.

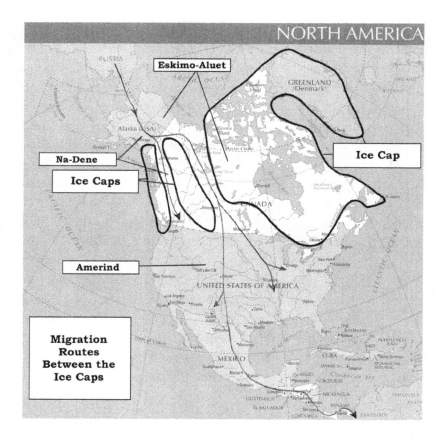

Man's Migration to the Western Hemisphere

28,000 BC (?) – 8,000 BC

Principal Language Groups:

Amerind, Na-Dene, Eskimo-Aluet

Figure 14

Amerind were the first to arrive with the Eskimo-Aleuts arriving last. Genetic differences including dominant blood type between Eskimos and American Indians also support at least two waves of immigrants to the North American Continent.[719]

Heretofore, this chapter has presented the conventional or traditional approach or consensus used in recent years to describe the prehistory of the hemisphere and the population that dwelt there. Charles C. Mann has written a provocative but well-researched and well-written book that examines the origins of the first peoples present in the Western Hemisphere. Mann presents evidence that portrays the Americas with a population greater than that of Europe. He describes Central American cities that were larger and finer than anything found elsewhere in the world and a civilization that was technologically advanced in the building of their cities and in the development of farming techniques. Mann asserts that the great majority of Indians lived south of the Rio Grande at the time of Columbus and that they were primarily farmers and not nomadic while others subsisted on fish and shellfish.[720] Mann also presents evidence that challenges the peopling of the Americas by those that crossed the Bering Strait. He quotes some authorities that suggest that Aborigines traveled from Australia by way of Antarctica to the southern tip of the Americas at Tierra del Fuego. Others propose that Paleo-Indians came down the Pacific Coast in boats that hugged the shoreline spotted with vegetation along the western edge of the vast ice sheet covering Alaska, Canada, and parts of the northern United States. One source speculated that such primitive boats could have traveled down the Pacific Coast of both the North and South American continents in as little as ten to fifteen years.[721]

Whatever the timetable and entry points, these early Americans were found in all parts of the hemisphere by 10,000 BC. This corresponds to significant climatic changes brought about by the end of the Pleistocene (glacial) geological epoch and the warming period thereafter. With the warming came continent-wide changes in prehistoric cultures by 6,000 BC. Vegetation increased and moved northward. Camels and mastodons died out and bison herds decreased. Some

PREHISTORY CULTURAL AREAS IN NORTH AMERICA.

Figure 15

food collectors began to farm. By 4,000 BC, only mountain glaciers remained.[722]

Many regions began to develop their basic characteristics by 6,000 BC. As people adapted to these regions, distinctive cultures arose. Three major cultural areas developed in the North American continent. The Eskimo cultural area included most of Alaska and stretched along the Northwest Territories of Canada. The Desert culture included the southwestern United States and the northwestern portion of Mexico. The Archaic culture covered a large area in the eastern half of the United States and southeastern Canada from Nova Scotia southwestward through the Great Lakes to near eastern Nebraska. From eastern Nebraska the area extended south to the Gulf coast and included all lands eastward to the Atlantic Ocean. Within the three large cultural zones (Eskimo, Desert, and Archaic) there existed various cultures and sub-cultures. Also, several subcultures existed outside of these three major cultural zones.[723]

With regard to eastern North America and its development in the prehistoric times following the glacial period, the Archaic culture began to emerge about 8,000 BC and ended in some places about 1,500 BC. In other areas the Archaic period lasted until the beginning of the historic period. The Archaic period was replaced, in part, by the Woodland or Mound Builder culture. This culture was substantially displaced by the Mississippian or Temple Builder culture between AD 500 and AD 1000. As with the Archaic period, the Woodland period remained in some areas until the emergence of the historical period. The Early and Middle Woodland culture have been labeled Burial Mound I and Burial Mound II cultures. The remaining Late Woodland culture is considered to be simultaneous with the Mississippian culture. The Mississippian culture is divided into Early and Late or Temple Mound I and Temple Mound II.[724]

PREHISTORY OF MAN IN EASTERN NORTH AMERICA.

1. Glacial Era: 28000 BC (?) – 8000 BC
2. Archaic Culture: 8000 BC – 1500 BC
A. Laurentian Culture (Northern)
B. Indian Knoll Culture (Southern)
3. Woodland Culture (Burial Mound I & II): 1500 BC – AD 500/ AD 1000
4. Misissippian Culture (Temple Mound I & II): AD 500/AD 1000-AD1500
5. End of Prehistory in North America: AD 1500

The suggested beginnings and endings of the various periods described above represent a distillation of information from several sources, each varying somewhat from the other. This table is merely an attempt to give some structure and sequence to the various periods of man's prehistory in North America.

By its very nature, knowledge of prehistory is, at best, guesswork based on archeological findings and the suppositions of scientists from various disciplines. As noted, the inhabitants of various geographical regions within a cultural area did not all evolve into a succeeding culture at the same time. There was considerable overlap of cultural periods with some cultures remaining unchanged long after others of a similar culture had been transformed into a succeeding culture.

Figure 16

The Archaic Period—8,000 BC to 1,500 BC

The Archaic culture's most notable achievements included wood-working and food-grinding tools; the greater use of land, rivers, and coasts; and the decline in movement from one region to another as the culture developed a greater dependence on local resources.[725]

The Archaic culture is divided into two major segments. A people with Algonquin linguistic traits sparsely occupied the northern woodlands and forests of the eastern United States and southeastern Canada. The Algonquins stretched from Labrador to the area surrounding the Great Lakes and back eastward to the Atlantic Ocean. A few even found their way into Virginia and the Carolinas. Typically, the Algonquins were tall with rather long heads and narrow faces. During Archaic times they were widely disbursed in small groups that hunted game, fished, and gathered berries, nuts, acorns, and roots. The French had known of the Algonquins since the early 1500s and described them as a gloomy, stern-faced people. Life for the Algonquin was a constant struggle as they battled starvation each winter. The Algonquins of the Archaic period viewed the world as being full of spirits, and their religion was based on fear and the power of evil.[726] The northern portion of the Archaic cultural area occupied by the Algonquins is known as the Laurentian culture.

The southern half of the Archaic cultural region is called the Indian Knoll culture. The Indian Knoll culture is the older of the two and includes the Carolina piedmont, the Florida coast, the lower Mississippi valley, and the uplands of the Tennessee and the Ohio River valleys. Large rock shelters and village houses were used as dwelling places and are a distinctive characteristic of the Indian Knoll culture. Houses were built with upright posts that supported a roof of perishable fibers. Floors were covered with clay and included a hearth.[727]

The people of the Indian Knoll culture were substantially if not totally Siouan. They were of medium stature and had high and vaulted heads, short faces, and medium broad noses. They have been called the Iswanids, meaning river people. Initially, the Siouans wandered about and lived by hunting and fishing. Because of the abundance of shellfish on the river banks, the formation of permanent villages at these locations resulted as shellfish became the main staple of the Siouan diet. No longer wanderers, the Siouans developed considerable piles of refuse in their permanent villages. The shells from shellfish and other village wastes were piled on the ground, and these piles eventually became large mounds or knolls. In time villages were established on top of these mounds, and the Siouans began burying their dead in pits dug in the tops of the mounds. The mounds ranged from four to twelve feet high and cover areas as large as seventeen acres. Such villages were first found in northern Kentucky, southern Indiana, and Ohio.[728]

The Woodland Period—1500 BC - AD 500/AD 1000

The Early Woodland period brought significant changes to the Indian Knoll culture. This was evidenced by the use of earthenware pottery, the growing of crops, and the emergence of a new religion that sought new gods in the heavens as opposed to the worship of earth-bound animal gods. The catalyst for many of these changes resulted from the migration to the Ohio Valley of a group of Indians from the south. The Adena[729] were relatively small, roundheaded, and broad faced. Into the Ohio Valley they brought their new faith and the practice of burying their dead.[730] During the Woodland period, villages grew larger and more numerous and were the forerunners of emerging tribal entities.

The northerners in that part of Ohio where the Adena settled readily accepted the Adena culture brought from the south. The northerners, outnumbering the Adena three to one, merely adopted the culture and soon took the leadership in bringing the culture into full flower. The cultural ascendance of the northerners over the Adena is known as the Mound Builder or Hopewell period. The Adena people were not driven out, but their faith and practices were absorbed during the Hopewell period.[731] The most visible evidence of the Mound Builders' culture is the ceremonial centers with great earth-walled enclosures. The earth-walled enclosures took the shape of circles, squares, rectangles, ovals, octagons, crescents, and various combinations of these shapes. Most structures embraced less than five acres, but some were as large as fifty acres. One such enclosure at Newark in Licking County, Ohio, is a circle that encloses thirty acres. Its wall is still five to fourteen feet high with a base of thirty-five to fifty-five feet. Inside and parallel to the wall is a ditch twenty-eight to forty-one feet wide and eight to thirteen feet deep.[732] Although remaining barbaric in many respects, the fine material culture of the later Mound Builder period and general well-being of the people indicate a significant advance of civilization.[733]

Some writers believe that the Kentucky Indian Knoll people were the earliest Indian group whose migrations can be tracked with a reasonable degree of assurance. Their shell mound culture spread in every direction from their old central location. They appear to have been a relatively large and homogeneous group with one language. As smaller groups formed, separated themselves from their Indian Knoll family, and migrated away from the old central location, many Siouan tribal groups were formed.[734]

The Mound Builder faith and culture spread rapidly in many directions, most strongly into Kentucky, Tennessee, Indiana, Illinois, and Wisconsin. Other, weaker penetrations were made into Missouri, Iowa, and Minnesota. It is generally assumed that other Siouans in Kentucky, Tennessee, and Illinois were attracted by the new culture and moved north of the Ohio River. Archaeological evidence strongly indicates that the Mound Builders were Siouan and in friendly contact with other Indians of the same stock. It is probable that traces of the Siouan groups that included the Quapaw, Osage, Omaha, Kansa, and Ponca tribes (the Dhegiha Sioux) were found in the southern part of Indiana. Based on the Dhegiha Sioux traditions of a former residence on the Ohio or Wabash rivers, they were most likely there during the Mound Builder period. The Dhegiha Sioux traditions included burials of the Mound Builder type in southern Indiana.[735] In summary, two racial groups share responsibility for development of the Mound Builder culture: the northern group believed to be Siouan and the southern Adena group.[736]

Mississippian Period—AD 500/AD 1000

The Hopewell culture declined about AD 300. The Mississippian or Temple Builder culture began to assert itself between AD 500 and AD 1000 and featured influences from Mexico and Middle America (the territories controlled by the Aztecs and the Maya in Mesoamerica). The Mississippian culture appears to have developed along the Red River where Texas, Oklahoma, Arkansas, and Louisiana meet. This area stands at the northern end of an overland trade route stretching northward through Mexico and Texas.[737] Most of the Indians of this area appear to have been Tunican, Natchezan, or other groups who were sun worshipers that revered the Sun Kings. But the Muskhogean people from lands west of the Mississippi were the driving force in introducing the Temple Builder or Mississippian culture eastward into central Georgia and northeastward from the mouth of the Ohio River at the Mississippi. This push up the Ohio River reached the Ohio valley along with another Muskhogean push from Tennessee. Another great center of Temple Builder influence was established in Illinois near present-day St. Louis. Although a war-like people, these penetrations by the Muskhogeans appear to have been peaceful. The Muskhogeans came among the Siouan Indians located in Illinois and southern Indiana. The Dhegiha Siouans (Quapaw, Ponca, Osage, Omaha, and Kansa tribes) were located in several districts along the Ohio and particularly in southern Indiana. Archaeologists indicate that the original Siouan Indians in Illinois were of the Kentucky Indian Knoll type and had been present in Illinois from the Archaic and Early Woodland periods to the end of the prehistoric period about AD 1500.[738] The arrival and conquest by the Europeans ended further development of the Temple Mound culture and the Mesoamerican influence on it.[739]

Historians with an interest in North America owe a considerable debt to the 300 French Jesuit missionaries that carried their faith to the Indians of the eastern parts of Canada and the United States between 1611 and 1764. These members of the Society of Jesus would trek the vast wilderness, paddle its waterways, sit at the campfires of many a tribe and tongue, endure incredible hardships, and frequently lose their lives doing so. They came not only to convert but also to learn the Indians' ways. Because there were no cities to hint at past cultures, no palace or temple walls upon which great events of the past would be memorialized, and no depositories of written history, the Jesuits' knowledge of the Indian would come only from talking with and observing them. The task would be daunting as the Jesuits attempted to separate fable from fact. With meticulous care they recorded their observations as well as the stories, legends, myths, and folklore mined from the memories passed from generation to generation. Fortunately, the Jesuits would send back to their superiors in Quebec their letters, journals, and reports of their work and the history of those among whom they labored. These reports would be attached or incorporated into an annual report to church officials in Paris. For over a century and a half these reports flowed across the Atlantic. Reports for the years 1632–1673 were published and would become known as the *Jesuit Relations.* These and other unpublished documents formed the basis for much of what we know about the aboriginal inhabitants for the Jesuits were the first and best historians of North America. Their works stretched from Labrador and Hudson Bay down the east coast of Canada, across the Great Lakes to Wisconsin, south through Illinois, and down the Mississippi River to New Orleans. During their tenure in North America the black-robed priests would interact and minister to the Hurons, Iroquois, Abenakis, Montagnais, Ottawas, Illinois, Sioux, Foxes, Tchactas, Akensas (Quapaw), and others.[740]

Significant evidence suggests that the cultural heritage of the Quapaw began with those Ameriends who roamed the eastern half of the United States during the Archaic period. The Dhegiha Sioux, including the Quapaw, appear to have been a part of the Indian Knoll people centered in the Ohio Valley who progressed from the Archaic period through the Mound Builder (Woodland) and Temple Mound (Mississippian) periods to historic times.

Near the beginning of the historic period, something caused many tribes including the Quapaw and other Dhegiha Sioux to abandon the Ohio Valley, which their ancestors had occupied for millennia. To determine the cause, we must turn our attention to the Iroquois.

The Iroquois are related to the Caddoans and Pawnees by blood and language. The timing of the Iroquois's separation from the Caddoans and migration from lands west of the Mississippi River northward into Western Pennsylvania, New York, and Ontario is not known.[741] However, there is some conjecture that the Iroquois arrived in the Algonquin areas of the northeastern United States about 1440. Over time the Iroquois tribes grew in population and rivaled the native Algonquins. Anthropologists have divided the Iroquois Proper into two groups. The Mohawks, Onondagas, and Oneidas comprised the eastern group. Senecas and Cayugas formed the western group. Other tribes were related to the Iroquois by blood and language: Hurons, Tobacco Nation (Tionotatis), Eries, Neutrals, and the Andastes.[742] The Iroquois's material culture was much higher than that of the nomadic Algonquin hunters and fishermen and included growing tobacco and crops and constructing villages.[743]

The Iroquois would eventually occupy central New York from the Hudson River on the east through the Finger Lakes region to the Genesee River on the west. Their lands would be surrounded like a giant island in a sea of Algonquin tribes. Algonquin-speaking tribes or dialects occupied Wisconsin, Michigan, Illinois, and Indiana while some hunting parties roamed Kentucky. Other Algonquins occupied parts of Virginia, Pennsylvania, New Jersey, southeastern New York, New England, New Brunswick, Nova Scotia, and the lower part of eastern Canada. New England was the most populous Algonquin region and included the Mohicans, Pequots, Narragansetts, Wampanoags, Massachusetts, and Penacooks.[744]

The other Iroquoian-speaking tribes would be found around the Great Lakes. The Hurons were located between the southeastern edge of Lake Huron and Lake Simcoe. The Eries resided below Lake Erie, and the Neutrals lived between the northern shore of Lake Erie and southern shore of Lake Ontario. The Andastes were located south of the Iroquois along the Susquehanna River in southern New York.[745]

The early Iroquoian migrants to the northeastern United States experienced defeat at the hands of the Algonquins along the St. Lawrence and were forced back to the lake country of central and northwestern New York sometime before 1535.[746] The French favored the Algonquins and the Hurons with trade. Although related to the Iroquois by blood and language, the Hurons were bitter enemies of the Iroquois because of long-standing quarrels stretching back into pre-historic times. The weapons of the Iroquois were no match for the steel knives and hatchets of their enemies obtained through trade with the French. Not only were they fighting their Huron cousins and the Algonquins, the Iroquois were fighting among themselves. However, the fortunes of war began to change for the Iroquois in 1570 when the five Iroquoian nations (Iroquois Proper) joined themselves and put their power in the hands of federal chiefs.[747] The fiercest of the five nations were the Mohawks, whose name means "cannibals."[748] The practices of torture and cannibalism were common among many Indians groups, but the Iroquois were unsurpassed in these areas. The Iroquois habitually practiced cannibalism after torturing their captives (men, women, and children) to death. The Iroquois took delight in feasting on the flesh of their victims.[749] The Iroquois confederacy was established as a means of conquering other Indian groups, not to battle any perceived threat posed by the few traders and white settlers at the eastern fringe of the continent. Europeon influence on the North American continent was minimal at best and is substantiated by the fact that what is now the oldest city in the United States was established in St. Augustine, Florida, in September 1565, only five years before the Iroquois confederacy was formed.[750]

Slaughter, torture, and cannibalism were not exclusive to the Iroquois five-nation confederacy. The Ottawas and Neutrals were at war with the various Michigan Algonquins. In the summer of 1643, the main stockade of the Nation du Feu defended by 900 warriors was attacked by the Neutrals. After a ten-day siege, the walls were breached and the Neutrals slaughtered part of the defenders, burned seventy of the warriors at the stake, and captured about 800 women, children, and old men. The old people were blinded, had their lips cut away, and then were released into the forest to starve or be killed by wild animals. The remaining women and children were led into captivity.[751]

The Winnebagoes were also cannibals and as ruthless and evil as the Iroquois confederacy. The Ilinis were an Algonquin tribe who befriended the Winnebagoes when the latter tribe was in a bad way. Five hundred Illini men took food to the hurting Winnebagoes. Supposedly, while a great feast was being held by the Illinis and Winnebagoes, the Winnebagoes cut the bowstrings of the Illinis and then suddenly attacked and killed their unsuspecting guests. The Winnebagoes followed the slaughter by cooking and eating the Illinis.[752]

One passage from the *Jesuit Relations* gives a general description of the treatment of captured enemies by the various Indian tribes of the era:

> In battle they strive especially to capture their enemies alive. Those who have been captured and led off to their villages are first stripped of their clothing; then they savagely tear off their nails one by one with their teeth; then they bind them to stakes and beat them as long as they please. Next they release them from their bonds, and compel them to pass back and forth between a double row of men armed with thorns, clubs, and instruments of iron. Finally, they kindle a fire about them, and roast the flesh of the muscles with red-hot plates and with spits, or cut it off and devour it, half-burned and dripping with gore and blood. Next, they plant blazing torches all over the body and especially in the gaping wounds; then, after scalping him they scatter ashes and live coals upon his naked head; then they tear the tendons of the arms and legs, lacerate them, or, after removing a little of the skin, leisurely cut them with a knife at the ankle and wrist. Often they compel the unhappy prisoner to walk through fire, or to eat and thus entomb in a living sepulcher, pieces of his own flesh.... Moreover, they prolong this torment throughout many days, and, in order that the poor victim may undergo fresh trials, intermit it for some time, until his vitality is entirely exhausted and he perishes.[753]

By 1600 the Iroquois were still poorly armed in comparison to their enemies. But by 1609, the Iroquois had a new source for weapons. The Dutch arrived at the Hudson and began to sell firearms as well as steel knives and hatchets to the Iroquois. During the first quarter of the seventeenth century, the Iroquois began to put their plans for conquest and destruction of their enemies into action.[754] The seeds of destruction for the Indians of the eastern United States during the next 150 years had been sown by the Iroquois. The white man's weapons may have made the Iroquois more efficient in their conquests, but the plan and execution of it was totally Indian. It must be remembered that the first permanent English settlement at Jamestown was not established until 1607. It was not until 1620 that the Pilgrims arrived on the shores of Massachusetts. Other tiny colonies followed. The devastation of the culture of the Indians of the Ohio Valley and other regions was substantially an Indian enterprise. It would be decades before the tiny, half-starved groups of Europeans that clung to the east coast of the United States would grow to sufficient size and strength to have a bearing on the demise of the Indian cultures. By that time, it was too late.

George Hyde in his book *Indians of the Woodlands from Prehistoric Times to 1725* called the Iroquois "...the Hitler Nazis of the Indian world. They had the same haughty faith in their being a master race as the German fanatics, the same cold-blooded pitilessness in dealing with weaker peoples."[755] Many examples serve to illustrate the ruthlessness of the Iroquois. In July 1648, the Iroquois attacked three fortified Huron towns, slaughtered the people, and took hundreds of women and children as slaves. In March 1649, a thousand Iroquois composed mostly of Mohawks and Senecas attacked three large Huron villages in their continued plan to destroy the Huron Nation. Huron survivors fled to the Tobacco Nation and soon the Iroquois pursued. After luring the Tobacco Nation warriors away from their stockade, the Iroquois attacked, threw firebrands into the wooden structures, and plundered the town. Many old people and women were killed, and children were thrown into burning houses. The remaining young women and children were taken captive and marched into the forest. Those that could not keep up the swift pace were killed. In the eighteen months between July 1648 and December 1649, the Hurons and the Tobacco Nation were effectively destroyed by the Iroquois.[756] The Iroquois then turned their attention to destruction of the Neutrals in 1650–1651 and the Eries in 1653–1656. As with the Hurons and Tobacco Nation Hurons, the Neutrals and Eries were all nations of the same blood and language as the Iroquois five-nation confederacy.[757] Only one enemy of their kindred remained, the numerically inferior Andastes. The Andastes war of attrition with the Iroquois confederation began with the Mohawks in 1650 and lasted until 1675 when the Senecas would finally defeat the brave and audacious Andastes. However, the cost to the victors was high as Iroquois towns had become depopulated, and many replacements were not of the Iroquoian stock but were adopted prisoners from many of the conquered nations.[758]

As their native cultures began to wither away, "…an Iroquois chief said with bitterness, the white man took the Indian by the hand in pretended friendship and then threw him behind him."[759] Yet, this is the very tactic the Iroquois used to defeat and destroy many other tribes, both before and after the arrival of the Europeans.

However, the spread of communicable diseases had a far greater impact than the Indian wars on the American Indian population during the 200 years following the arrival of the Europeans. The precipitous decline of the North American Indian population began almost immediately after the first European arrivals in the Caribbean. The diseases spread rapidly through North America years before the first settlers set foot on the continent. The Europeans had developed significant resistance to the diseases through repeated exposure and natural selection.[760] However, the Indians of the Americas, insulated for several millennia by two oceans, had not been exposed to smallpox, cholera, bubonic plague, influenza, typhus, typhoid, measles, and scarlet fever. With almost a total lack of resistance to the various diseases, mortality rates were devastatingly high.[761]

The magnitude of the loss of Indian life can only be understood by examination of population figures. Only during the past one hundred years has serious study been given to the size of the Indian population in North America north of the Rio Grande in 1492. The first scholarly effort by James Mooney was published in 1910 and estimated a North American Indian population of slightly over one million at the time of Columbus' arrival.[762] Applying various techniques to archaeological data and the early post-Columbian written records of a few literate Europeans, demographers now generally agree on that the Indian population ranged between two million and five million, but other estimates were as high as eighteen million.[763] It is believed that the most devastating losses occurred in the first 100 to 200 years. By 1700, many tribes were extinct or almost so.[764] Henry Schoolcraft, under the direction of the Bureau of Indian Affairs, conducted a survey of the Indian population between 1847 and 1850. Schoolcraft calculated the population of each tribe and arrived at a cumulative population figure of approximately 388,000 plus an estimated 25,000 to 35,000 Indians within the unexplored territories of the United States.[765] Whatever the number, the dramatic population decline began almost immediately and continued to its lowest point in 1890 when the estimated North American Indian population was 228,000.[766]

There are two predominant non-Indian views of the American Indian. One view holds that the Indians are morally and intellectually inferior. The second view is held by those persons who have a particularly anti-Western culture mindset and ascribe to the Indian all of those traits and characteristics that are opposites of what they perceive to be the evils of a white society—corruption, greed, treachery, untrustworthiness, spoilers of nature, and spiritually out-of-tune with the rhythms of the universe. This group views the Indian as noble savages who have been victimized by Western civilization. This view has grown during the last half of the twentieth century with academia as its main source. This group owes its views more to a hatred of Western culture than to a proper understanding of the American Indian.[767] Therefore, it is often the goal of modern-day historians to romanticize the Indian, blame the arrival of the European for all of his ills, bathe the Indian in nobility, and idealize his culture. As one writer has stated with regard to the Quapaws,

> ...there is not the first indication in the mass of surviving letters and other primary sources that bear on the history of colonial Arkansas that the Quapaws were environmentalists or conservationists, or that they lived in some magically balanced communion with nature before the arrival of Europeans somehow upset the delicate prelapsarian harmony.[768]

As with most cultures, the truth about the American Indian lies somewhere between these opposing views.

Certainly, the Indian did not fare well as the two societies clashed. But the savagery and butchery between Indian groups were well established centuries before the arrival of the white man. Other than the sporadic forays into the interior of the continent by a handful of Europeans during the sixteenth century, the white man had little contact with the Indians prior to the 1600s. The attempts to sanitize the history of Indian society prior to the Europeans' arrival and extensive contact with the Indians are illustrated by one writer's comments.

> White men publicized images of the Indians as savage and "bestial" warriors, but Indian warfare was often not more savage than the type of war Europeans introduced to the hemisphere and waged against the natives. In fact, Indian warfare was often stimulated and intensified by the impact of the Europeans.[769]

Given the preponderance of evidence of widespread cannibalism; torture of captives; slavery; indiscriminate killing of the old people, women, and children; this writer's statement holds little credibility. This writer continues by stating, "Warfare, even after the arrival of the whites, usually consisted of sporadic raids, conducted in defense of tribal lands and hunting grounds, or for honor, revenge, slaves, horses, or other booty. There were few sieges, protracted battles or wars of conquest."[770] However, it was white man and the ravages of disease that ultimately brought a stop to the inter-tribal warfare that raged through most of the sixteenth and seventeenth centuries and into the beginning of the eighteenth century.

There is not a specific date for the beginning of the Iroquoian wars on the Ohio Valley Indians tribes. The Iroquois campaign against the Eries occurred between 1653 and 1656 with some of the Eries retreating to the Ohio River. This may have brought the Iroquois quest for destruction and domination to the Ohio Valley about 1655. By the time the French came to the Mississippi in 1673, the Iroquois had driven the Dhegiha Sioux to the Mississippi River and beyond.[771] The Dhegiha Sioux moved down the Ohio to the Mississippi in separate groups and most likely occurred between these initial Iroquois attacks in the Ohio Valley in the 1650s and the initial French contact with the Quapaws in Arkansas in 1673. The Omahas and Poncas fled into northwestern Iowa, the Kansa fled to eastern Kansas, and the Osages to the Osage River, all westward through Missouri from the confluence of the Ohio with the Mississippi. However, the Quapaw fled down the Mississippi along the east bank, and forever after they would be known as the Downstream People.[772]

BIBLIOGRAPHY

Books

Arnold, Morris S. *The Rumble of a Distant Drum*. Fayetteville, Arkansas: The University of Arkansas Press, 2000.

Aylesworth, Thomas G., and Virginia L. Aylesworth. *Washington—The Nation's Capitol*. New York: Smithmark Publishers, Inc., 1991.

Baird, W. David. *The Quapaws*. New York: Chelsea House Publishers, 1989.

_____. *The Quapaw Indians—A History of the Downstream People*. Norman, Oklahoma: University of Oklahoma Press, 1980.

Bennett, William J. *The Broken Hearth*. New York: Doubleday, 2001.

Blosser, Howard W. *Prairie Jackpot*. Webb City, Missouri, 1973.

Bossu, Jean-Bernard. *New Travels in North America*, Samuel Dorris Dickinson, trans., ed. Natchitoches, Louisiana: Northwestern State University Press, 1982.

Burke, James. *The Pinball Effect*. Boston: Little, Brown & Co., 1996.

Burleigh, Nina. *The Stranger and the Statesman*. New York: William Morrow, 2003.

Burrough, Bryan. *Public Enemies*. New York: The Penguin Press, 2004.

Champagne, Duane, ed. *The Native North American Almanac*. Detroit, Michigan: Gale Research, Inc. 1994.

Clayman, Charles B., MD, ed. "Pneumoconiosis." *Home Medical Encyclopedia*. Vol. Two—I-Z. New York: Random House, 1989.

Commager, Henry Steele, ed. *Documents of American History*. Vol. 1 to 1865. New York: F. S. Crofts & Co., 1935.

_____. *Documents of American History*. Vol. II since 1865. New York: F. S. Crofts & Co., 1935.

Driver, Harold E. *Indians of North America*. Second Edition, Revised. Chicago: The University of Chicago Press, 1969.

Ellis, John. "The War in the Trenches: Guns and Gas." *World War I*. Donald J. Murphy, ed. San Diego, California: Greenhaven Press, Inc., 2002.

Faulk, Odie B. *Oklahoma—Land of the Fair God.* Northridge, California: Windsor Publications, Inc., 1986.

Foreman, Grant, ed. and anno. *A Traveler in Indian Territory.* Norman, Oklahoma, University of Oklahoma Press, 1930.

_____. *Advancing the Frontier.* Norman, Oklahoma, University of Oklahoma Press, 1933.

Galena Bicentennial Committee. *Pioneer Days of Galena.* Galena, Kansas: Tri-State Printing Co., 1984.

Galloway, Patricia. "Couture, Tonti, and the English-Quapaw Connection, A Revision." Arkansas

Archeological Survey Research Series No. 40. *Arkansas Before the Americans.* Hester A. Davis, ed. Wrightsville, Arkansas: Arkansas Department of Corrections, 1991.

Gibson, Arrell Morgan. *Oklahoma—A History of Five Centuries,* Second Edition. Norman, Oklahoma: University of Oklahoma Press, 1981.

_____. *Wilderness Bonanza.* Norman, Oklahoma: University of Oklahoma Press, 1972.

Gorenstein, Shirley, ed. *North America.* New York: St. Martin's Press, 1975.

Hartcup, Guy. *The War of Invention.* London: Brassey's Defence Publishers, 1988.

Hoig, Stan. *The Oklahoma Land Rush of 1889.* Oklahoma City, Oklahoma: Oklahoma Historical Society, 1984.

Hogg, Ian V. *Allied Artillery of World War One.* Wiltshire, England: The Crowood Press, 1998.

Hyde, George E. *Indians of the Woodlands—From Prehistoric Times to 1725.* Norman, Oklahoma: University of Oklahoma Press, 1962.

Jackson, Kevin, and Jonathan Stamp. *Building the Great Pyramid.* London: Firefly Books, 2002.

Josephy, Alvin M., Jr. *The Indian Heritage of America.* New York: Alfred A. Knopf, 1990.

Kenton, Edna, ed. *The Indians of North America,* Vol I & II. New York: Harcourt, Brace & Company, 1927.

Kurten, Bjorn. *Before the Indians.* New York: Columbia University Press, 1988.

Lewis, Anna. *Along the Arkansas.* Dallas: The Southwest Press, 1932.

Mann, Charles C. *1491.* New York: Alfred A. Knopf, 2005.

Mantle, Merlyn, Mickey Mantle, Jr., David Mantle, and Dan Mantle with Mickey Herskowitz. *A Hero All His Life.* New York: Harper Collins Publishers, 1996.

Mantle, Mickey, Herb Gluck, *The Mick.* Garden City, New York: Doubleday & Company, Inc., 1985.

McCullough, David. *John Adams.* New York: Simon & Schuster, 2001.

McGuire, Lloyd H., Jr. *Birth of Guthrie.* San Diego, California: Crest Offset Printing Company, 1998.

Morgan, Anne Hodges, and H. Wayne Morgan, Ed. *Oklahoma—New Views of the Forty-Sixth State.* Norman, Oklahoma: University of Oklahoma Press, 1982.

Morgan, H. Wayne, and Anne Hodges Morgan. *Oklahoma—A History.* New York: W. W. Norton & Company, 1984.

Morris, Edmund. *Theodore Rex.* New York: The Modern Library, 2001.

Morris, Richard B., ed. "The Civil War and Reconstruction, 1861–1877." *Encyclopedia of American History.* New York: Harper Brothers, 1953.

_____. "The Colonies and the Empire, 1624–1775." *Encyclopedia of American History.* New York: Harper & Brothers, 1953.

_____. "The Early National and Ante-Bellum Periods, 1789–1860." *Encyclopedia of American History.* New York: Harper & Brothers, 1953.

_____. "Founding of the English Colonies, 1578–1732." *Encyclopedia of American History.* New York: Harper & Brothers, 1953.

Morse, Dan F. "On the Possible Origin of the Quapaws in Northeast Arkansas." Arkansas Archeological Survey Research Series No. 40. *Arkansas Before the Americans.* Hester A. Davis, ed. Wrightsville, Arkansas: Arkansas Department of Corrections, 1991.

Murphy, Donald J., ed. *World War I.* San Diego, California: Greenhaven Press, Inc., 2002.

Newsom, D. Earl. *The Cherokee Strip—Its History & Grand Opening.* Stillwater, Oklahoma: New Forums Press, Inc., 1992.

Nieberding, Velma. *The History of Ottawa County.* Marceline, Missouri: Walsworth Pub. Co., 1983.

_____. *The Quapaws (Those who went downstream).* Miami, Oklahoma: Dixons, Inc., 1976.

Nuttall, Thomas. *A Journal of Travels into the Arkansas Territory during the year 1819.* Savoie Lottinville, ed. Norman, Oklahoma: University of Oklahoma Press, 1980.

Parkman, Francis. *The Jesuits in North America in the Seventeenth Century*. Boston: Little, Brown, and Company, 1905.

Probst, Katherine N., Don Fullerton, Robert E. Litan, and Paul R. Portney. *Footing the Bill for Superfund Cleanups*. Washington, D.C.: The Brookings Institution and Resources for the Future, 1995.

Ross, Malcolm H. *Death of a Yale Man*. New York: Farrar & Rinehart, Inc, 1939.

Ryan, Frederick Lynn. *Problems of the Oklahoma Labor Market, with Special Reference to Unemployment Compensation*. Oklahoma City, Oklahoma: Semco Color Press, Inc., 1937.

Sabo, George, III. "Inconsistent Kin: French-Quapaw Relations at Arkansas Post." Arkansas Archeological Survey Research Series No. 40. *Arkansas Before the Americans*. Hester A. Davis, ed. Wrightsville, Arkansas: Arkansas Department of Corrections, 1991.

Schoolcraft, Henry R. *History, Condition and Prospects of the Indian Tribes of the United States*. Philadelphia: Lippincott, Grambo & Company, 1853.

Shaner, Dolph. *The Story of Joplin*. New York: Stratford House, Inc., 1948.

Snipp, C. Matthew. *American Indians: The First of This Land*. New York: Russell Sage Foundation, 1989.

Starr, Paul. *The Creation of the Media*. New York: Basic Books, 2004.

Stewart, Dr. David, and Dr. Ray Knox. *The Earthquake That Never Went Away*. Marble Hill, Missouri: Gutenberg-Richter Publications, 1993.

Strickland, Rennard. *The Indians of Oklahoma*. Norman, Oklahoma: University of Oklahoma Press, 1980.

Suggs, George G., Jr. *Union Busting in the Tri-State*. Norman, Oklahoma: University of Oklahoma Press, 1986.

Swanson, Earl H., Warwick Bray, and Ian Farrington. *The Ancient Americas*. New York: Peter Bedrick Books, 1989.

Taylor, F. Jay, *The United States and the Spanish Civil War*. New York: Bookman Associates, 1956.

Thoburn, Joseph B. and Muriel H. Wright. *Oklahoma—A History of the State and Its People*. Volume II. New York: Lewis Historical Publishing Company, Inc., 1929.

Thompson, C. J. S. *The Lure and Romance of Alchemy*. New York: Bell Publishing Company, 1990.

Thompson, Vern E. *Brief History of the Quapaw Tribe of Indians*. Pittsburg, Kansas: Mostly Books, Originally published 1937.

Tocqueville, Alexis de. *Democracy in America*. The Henry Reeve Text as Revised by Francis Bowen, Volume I. New York, Alfred A Knoff, 1945.

_____. *Democracy in America*. Revised Edition, Vol. II, Second Book. London: The Colonial Press, 1900.

Toland, John. *Dillinger Days*. New York: Random House, 1963; New York: Da Capo Press, Inc., 1995. Citation is to the Da Capo edition.

Tucker, Spencer C. *The Great War 1914–18*. Bloomington & Indianapolis: Indiana University Press, 1998.

Wallis, Michael. *Oil Man*. New York: Doubleday, 1988.

Watt, Susan. *Lead*. New York, New York: Benchmark Books, 2002.

Weidman, Samuel. *Miami-Picher Zinc-Lead District*. Norman, Oklahoma: University of Oklahoma Press, 1932.

World Almanac and Book of Facts 1999. Mahwah, New Jersey: World Almanac Books, 1998.

Periodicals

Dale, Edward Everett. "Two Mississippi Valley Frontiers." *Chronicles of Oklahoma*. Vol. XXVI. Winter 1948–49—382.

Foreman, Carolyn Thomas. "Education Among the Quapaws." *The Chronicles of Oklahoma*. Vol. XXV. Spring 1947—16–17, 22, 25.

_____. "Education Among the Quapaws 1829–1875." *The Chronicles of Oklahoma*. Vol. XLII. Summer 1964—15.

_____. "Lewis Francis Hadley: The Long-Haired Sign Talker." *The Chronicles of Oklahoma*. Vol. XXVII. Spring 1949—42, 47.

Gibson, Arrell Morgan. "Poor Man's Camp." *The Chronicles of Oklahoma*. Vol. LX. Spring 1982—10- 14.

Harbaugh, M. D. "Labor Relations in the Tri-State Mining District." *Mining Congress Journal*. 22. June 1936—19–21, 24.

Harris, Frank H. "Neosho Agency 1838–1871." *The Chronicles of Oklahoma*. Vol. XLIII. Spring, 1965—54–55.

Harriss, Joseph. "Westward Ho!" *Smithsonian*. Vol. 34, Number 1. April 2003—104.

Jackson, Joe C. "Schools Among the Minor Tribes in Indian Territory." *The Chronicles of Oklahoma*. Vol. XXXII. 1954—58.

Jones, Landon Y. "Iron Will." *Smithsonian*. Vol. 33, Number 5. August 2002—98.

"Joplin—Dec.2." *Engineering and Mining Journal.* Vol. XCVIII. December 5, 1914—1062.

Parker, Linda. "Indian Colonization in Northeastern and Central Indian Territory." *The Chronicles of Oklahoma.* Vol. LIV. Spring 1976—108.

"Teacher Supplement to the Arkansas News." Old State House Museum. *The Arkansas News Teachers Guide.* Spring 1988—2.

Tobin, James. "To Fly!" *Smithsonian.* Vol. 34, Number 1. April 2003—61–62.

Thompson, Vern E. "A History of the Quapaw." *The Chronicles of Oklahoma.* Vol. XXXIII. 1955—360, 366–367, 369, 395–397.

"Tri-State Strike Broken." News of the Industry. *Engineering and Mining Journal.* Vol. CXXXVI. July 1935—347.

Newspapers

Averill, David, "Dodging the bullet: Abandoned mines ongoing hazard," *Tulsa World,* March 14, 2004.

_____, "Hellholes," *Tulsa World,* June 22, 2003.

Brookshire, Amerillis, "Tobacco chewing mule," *Tri-State Tribune,* June 12, 2003.

Chestnut, Charles, "Those days in the mines," *Tri-State Tribune,* August 9, 1990.

Daily Oklahoman, "Mine Town Nurse," November 21, 1948.

DelCour, Julie, "The town that Jack built," *Tulsa World,* June 22, 2003.

Gillham, Omer, "Collapsing into the past," *Tulsa World,* February 19, 2006.

Jones, Mike, "That sinking feeling," *Tulsa World,* November 30, 2003.

Joplin Globe, "Accident Fatal to Rufus Nolan," April 28, 1925.

_____. "Baxter tailing mill operator injured in attack by two men," May 21, 1935.

_____. "Four miners plunge to their deaths in a mine shaft," October 5, 1928.

_____. "Militia reaches mine field after day of rioting," May 28, 1935.

_____. "1 dead, 3 hurt in mine explosion," August 5, 1928, 1.

_____. "Ten hurt, one fatally in mine gas explosion," November 28, 1925.

_____. "3000 at union miners' meeting," May 22, 1935.

Joplin News Herald, "Inquest tonight into death of John Hostetter," August 25, 1926.

_____. "Miner's fall down shaft fatal to two," December 9, 1925.

_____. "Second Picher mine accident victim dies," April 30, 1926.

Luthy, Brenda, "Kansas sinkhole continues to grow," *Tulsa World,* August 2, 2006.

Matthews, Allan, "Picher growing pains—the first dozen years," *Tri-State tribune,* August 9, 1990.

Miami (Oklahoma) Daily News-Record, "E-P suspending mine, mill operations across district," April 23, 1957.

_____. "Governor sends troops here to quell disorder," May 27, 1935.

_____. "Miner is wounded," June 9, 1935.

_____. "Miners to hold meeting today," May 26, 1935.

_____. "Picher Area Shocked by E-P News: Other firms to continue in the field," April 7, 1957.

_____. "Pollution solution," September 7, 1981.

_____. "230 guardsmen arrive to put lid on violence in Kansas mining field," June 9, 1935.

Miami News Record, "Discovery of man's body in mine shaft leads to detention of his uncle," January 18, 1925.

_____. "Miner seriously hurt by falling boulder," March 19, 1925.

_____. "Wilbur Kell," June 7, 1925.

Miami Record Herald, "Joe Allen killed-fell into shaft," March 27, 1914.

_____. "Man hurt in accident-dies here," July 21, 1927.

Moody, Ben, "O. W. (Ol) Sparks, a mining man," *Tri-State Tribune,* August 9, 1990.

_____. "Yester-year in the Picher mining field—Baseball in the Picher Mining Field," *Tri-State Tribune,* October 29, 1992.

_____. "Yester-year In The Picher Mining Field—Building a Central Mill," *Tri-State Tribune,* September 8, 1988.

_____. "Yester-year In The Picher Mining Field—The Building of a Central Mill," *Tri-State Tribune,* September 15, 1988.

_____. "Yester-year In The Picher Mining Field—The Building of The Central Mill," *Tri-State Tribune,* September 22, 1988.

_____. "Yester-year In The Picher Mining Field—The Building of The Central Mill," *Tri-State Tribune,* September 29, 1988.

_____. "Yester-year In The Picher Mining Field—The Building of The Central Mill (Continued)," *Tri-State Tribune*, October 6, 1988.

_____. "Yester-year In The Picher Mining Field—Central Mill Chat Loading," *Tri-State Tribune*, June 30, 1988.

_____. "Yester-year in the Picher mining field—Mickey Mantle," *Tri-State Tribune*, November 5, 1992.

_____. "Yester-year In The Picher Mining Field—The Miners' Hard Hat," *Tri-State Tribune*, February 9, 1989.

_____. "Yester-year In The Picher Mining Field—Silicosis," *Tri-State Tribune*, January 26, 1989.

Myers, Jim, "Inhofe to unveil Tar Creek Plan," *Tulsa World*, May 4, 2006.

_____. "Inhofe: U. S. 69 could cave in," *Tulsa World*, December 16, 2005.

_____. "Tar Creek plan pushed," *Tulsa World*, May 5, 2003.

Neal, Ken, "Beyond dispute," *Tulsa World*, February 19, 2006.

The New Republic, "American Plague Spot," Vol. CII, No. 1, January 1, 1940.

Patton, Marjorie Ann, "Women in the Mines," *The Chat Pile*, June 10, 1998.

Pearson, Janet, "Inhofe plan not the answer," *Tulsa World*, July 1, 2003.

Pulley, Betty, "The History of Picher," *The Chat Pile*, June 10, 1998.

Rentfrow, Marie, "The Town that Jack Built," *The Chat Pile*, June 10, 1998.

Ruckman, S. E., "Quapaw tradition lives on, even in death," *Tulsa World*, May 11, 2005.

Schafer, Shaun, "Ride to offer 'Toxic Tour'," *Tulsa World*, May 1, 2003.

_____. "Tar Creek buyout idea gets push," *Tulsa World*, May 3, 2003.

_____. "Toiling with tainted soil," *Tulsa World*, April 26, 2003.

Tri-State Tribune, "City of Picher born out of a happy accident," May 30, 2002.

_____. "The Story of Old Toby, the last of the lead mine mules," August 9, 1990.

Tulsa World, "In Crisis," Tulsa World, February 21, 2005.

Wood, Frank, "Kangaroo courts common in mines," *Tri-State Tribune*, May 30, 2002.

_____. "Language of the mines," *Tri-State Tribune*, June 12, 2003.

_____. "Moving Mules," *Tri-State Tribune*, August 9, 1990.

Government Documents

41st Congress, 3rd Session, House Report 39, VII-IX, 51–52.

42nd Congress, 2nd Session, House Executive Documents 276, 1–2.

51st Congress, 1st Session, House Report 559, 1–2.

52nd Congress, 1st Session, House Report 2040, 1–2

52nd Congress, 1st Session, Senate Report 615, 1–3.

53rd Congress, 3rd Session, *Congressional Record,* 1233.

Annual Report of the Commissioner of Indian Affairs to the Secretary of the Interior for the Year 1839–1840. Washington: J. Gideon, Jr., Printer, 1839, 154. (University of Tulsa, McFarlin Library, Microfilm M no. 00345).

Annual Report of the Commissioner of Indian Affairs to the Secretary of the Interior for the Year 1867. Washington: Government Printing Office, 1868—272.

Annual Report of the Commissioner of Indian Affairs to the Secretary of the Interior for the Year 1868. Washington: Government Printing Office, 1868—324.

Annual Report of the Commissioner of Indian Affairs to the Secretary of the Interior for the Year 1869. Washington: Government Printing Office, 1870—38, 380–381.

Annual Report of the Commissioner of Indian Affairs to the Secretary of the Interior for the Year 1872. Washington: Government Printing Office, 1872—243.

Annual Report of the Commissioner of Indian Affairs to the Secretary of the Interior for the Year 1873. Washington: Government Printing Office, 1874—213–224.

Annual Report of the Commissioner of Indian Affairs to the Secretary of the Interior for the Year 1874. Washington: Government Printing Office, 1874—228.

Annual Report of the Commissioner of Indian Affairs to the Secretary of the Interior for the Year 1875. Washington: Government Printing Office, 1875—281.

Annual Report of the Commissioner of Indian Affairs to the Secretary of the Interior for the Year 1876. Washington: Government Printing Office, 1876—57.

Annual Report of the Commissioner of Indian Affairs to the Secretary of the Interior for the Year 1877. Washington: Government Printing Office, 1877—103.

Annual Report of the Commissioner of Indian Affairs to the Secretary of the Interior for the Year 1878. Washington: Government Printing Office, 1878—65.

Annual Report of the Commissioner of Indian Affairs to the Secretary of the Interior for the Year 1879. Washington: Government Printing Office, 1865—75–76, 78.

Annual Report of the Commissioner of Indian Affairs to the Secretary of the Interior for the Year 1880. Washington: Government Printing Office, 1880—86.

Annual Report of the Commissioner of Indian Affairs to the Secretary of the Interior for the Year 1889. Washington: Government Printing Office, 1889—198.

Indian Affairs: Laws and Treaties. Vol. VI Laws. Compiled from February 10, 1939 to January 13, 1971. Washington: Government Printing Office, 1971—1193.

Kappler, Charles J., comp. & ed. *Indian Affairs: Laws & Treaties,* Vol. 1 Laws. Compiled December 1, 1902. Washington: Government Printing Office, 1904—566–567, 619–620.

_____. *Indian Affairs: Laws and Treaties.* Vol. II Laws. Washington: Government Printing Office, 1904—160–161, 210–211, 961–962.

_____. *Indian Affairs: Laws & Treaties,* Vol. III Laws. Compiled December 1, 1913. Washington: Government Printing Office, 1913—387.

_____. *Indian Affairs: Laws & Treaties.* Vol. IV Laws, Compiled March 4, 1927. Washington: Government Printing Office, 1904—316.

Keheley, Ed, and Mary Ann Pritchard. *Report to Governor Keating's Tar Creek Superfund Task Force by the Subsidence Committee.* July 21, 2000—13.

Luza, Kenneth V. "Stability problems associated with abandoned underground mines in the Picher Field, Northeastern Oklahoma." *Oklahoma Geological Survey Circular 88.* Norman, Oklahoma, The University of Oklahoma, 1986—13–16.

National Labor Relations Board. "Eagle-Picher Mining & Smelting Company, et al., Case No. C-73." *Decisions and Orders.* Vol. 16. October 16–31, 1939. Washington: United States Government Printing Office, 1940—728, 741, 744, 750, 758, 762–763.

_____. "N. L. R. B. v. Jones & Laughlin Steel Corp., 301 U.S. 1." April 12, 1937. *Court Decisions Relating to the National Labor Relations Act.* Vol. I. Cases decided before December 31, 1939. Washington, D. C.: Government Printing Office, 1944—333–334.

Sayers, R. R., F. V. Meriwether, A. J. Lanza, and W. W. Adams. "Silicosis and Tuberculosis Among Miners of the Tri-State District of Oklahoma, Kansas, and Missouri-I." *U.S. Department of Commerce Technical Paper 545.* Washington, D. C.: Government Printing Office, 1933—4–5.

State of Oklahoma. Department of Environmental Quality. *Oklahoma Plan for Tar Creek.* May 2003.

U.S. Army Corps of Engineers. Subsidence Evaluation Team. Picher Mining Field, Northeast Oklahoma Subsidence Evaluation Report. January 2006—7.5–7.6.

U.S. Congress. Senate. Committee on Indian Affairs. Testimony of Abner W. Abrams. March 5, 1892—17.

U.S. Department of Commerce. Bureau of Census. *Fifteenth Census of the United States: 1930.* Volume III: Part 1, Alabama-Missouri. Washington, D. C.: Government Printing Office, 1932—830, 1332, 1334.

_____. Bureau of Census. *Fifteenth Census of the United States: 1930.* Volume III, Part 2, Montana-Wyoming, 554.

_____. Bureau of the Census. *Indian Population in the United States and Alaska 1910.* Washington: Government Printing Office, 1915—20, 37.

U.S. Department of Labor. Bureau of Labor Statistics. *Wage in the Nonferrous-Metals Industry, June 1943.* Washington, D.C.: Government Printing Office, June 1943—7.

U.S. Environmental Protection Agency. "Tar Creek (Ottawa County) Oklahoma, EPA ID# OKD980629844." EPA Region 6, Congressional District 02. EPA Publication Date: March 5, 2003—1.

_____. "Tar Creek (Ottawa County) Oklahoma, EPA ID# OKD980629844." EPA Region 6, Congressional District 02. EPA Publication Date: January 10, 2006—1-3.

_____. "EPA Superfund Record of Decision: Tar Creek (Ottawa County). EPA ID: OKD980629844. OU 01 Ottawa County, OK." June 6, 1984—# CSS Current Site Status, #AE Alternative Solutions.

Archival Materials

Galena Mining and Historical Association, Galena, Kansas. Information retrieved and supplied by Carla Taylor Jordan.

Quick, D. E. "A Great Race." *The Kendall Collegian.* May 1906. McFarlin Library College Files, University of Tulsa, Tulsa Oklahoma.

Tulsa World. Archieves. http://www.tulsaworld.com/ArchiveSearch/Search/SearchFormNew.asp

Electronic Sources

Agency for Toxic Substances and Disease Registry. *Report to Congress—Tar Creek Superfund Site, Ottawa County, Oklahoma.* http://.atsdr.cdc.gov/sites/tarcreek/tarcreekreport-pl.html

A. L. A. Schechter Poultry Corporation et al. v. Unites States, 295 U.S. 495 (1935).

Batt, Tony. "National Indian Gaming Report: Tribal gaming take: $19.4 billion." *Los Vegas Review Journal.* July 14, 2005. http://www.reviewjournal.com/lvrj_home/2005/Jul-14-Thu-2005/business/2501358.html

"Bartlesville Sports History." http://myweb.cableone.net/gmeador/sports.htm

Brainy Encyclopedia. "Zinc." http://www.brainyencyclopedia.com/encyclopedia/z/zi/sinc.html

Christenson, Scott. "Contamination of Wells Completed in the Roubidoux Aquifer by Abandoned Zinc and Lead Mines, Ottawa County, Oklahoma." *U. S. Geological Survey Water-Resources Investigations Report 95–4150.* Abstract, Table 1, p. 6. http:// pubs.usgs.gov/wri/wri954150

Cramb, Alan W. "A Short History of Metals." Carnegie Mellon University. http:// neon.mems.cmu/ cramb/processing/history.html

Encyclopedia of North American Indians—Indian Territory. http://college.hmco.com/ history/readerscomp/naind/html/na_016600_indianterrit.htm

"Gambling." *Focus on Social Issues.* http://www.family.org/cforum/fosi/gambling/ gitus/

Grand River Dam Authority. "Grand Lake O' the Cherokees." http://www.grda. com/waer/ grand.htm

Jefferson Lab. "It's Elemental." http://education.jlab.org/itselemental/ele030.html

The Joplin Globe Online Edition. "Tar Creek-In Our View." February 23, 2002. http:// joplinglobe.com/archives/2002/020203/headline/Story4.html

Lambert, James L. "Mickey Mantle's 11th Hour Miracle." Agape Press, August 12, 2005. http://headlines.agapepress.org/archive/8/122005f.asp

Los Angeles Chinese Learning Center. "General Information of Zinc http://chinese-school.netfirms.com/Zinc-information.html

Henry Methvin—April 8, 1912-April 19, 1948. http://www.tmethvin.com/henry/

Meyer, Richard E. "The Tar Creek Time Bomb." *Los Angeles Times.* http://www.sci-ence.uwaterloo.ca/earth/waton/s902.html

Daniel Patrick Moynihan. http://www.infoplease.com/ce6/people/A0834298.html

"Moynihan on the family/Wilson on Marriage & Divorce." Interview of Daniel Patrick Moynihan by George Will on *This Week,* American Broadcasting Company, September 24, 2000. http://archives.his.com/smartmarriages/2000-September/ msg00047.html

National Indian Gaming Commission. "National Indian Gaming Commission Tribal Gaming Revenues." http://www.nigc.gov/nigc/tribes/trigamrev2004to2003.jsp

Nile of the New World. "The Natural Environment: The Delta and Its Resources." *Draft Heritage Study and Environmental Assessment.* Vol. 2. http://www.cr.nps.gov/ delta/dhsea.htm

Polk, Jonna. "Tar Creek is nation's largest Superfund site." *U. S. Army Corps of Engi-neers, Tulsa District, May 2005 Engineer Update.* http://www.hq.usace.mil/ccpa/pubs/ may05/STORY11.htm

Problem Gambling.com. http://problemgamblinglcom/faq.html

Probst, Katherine N. "Superfund at 25: What Remains to be Done." *Resources.* Fall 2005.http://www.rff.org/rff/News/Features/Superfund-at-25.cfm

Quapaw Website. "Town Structure." htttp://geocities.com/Athens/Aegean/1388/tow.html?20072

_____. "Government Organization Throughout Time." http://geocities.com/Athens/Aegean/1388/gov.html?20072

Sabo, George, III. "The Osage Indians." *Historic Native Americans in the Mississippi Valley.* Arkansas Archeological Survey. Revised July 2001. http://www.uark.edu/depts/contact/osage.html

Sierra Club. "Communities at risk: Ottowa [sic], Oklahoma." http://www.sierraclub.org/ communities/Oklahoma/

Smithsonian Institution. "James Smithson—Founder of the Smithsonian Institution." http://www.si.edu/archives/documents/smithson.htm

State of Oklahoma. Department of Environmental Quality. "Tar Creek." *Fact Sheets.* http://www.deq.ok.us/1pdnew/FactSheet/Old Remediation Reports/RemREp10-02%.pdf.

TEF Enterprises. "Sixteen Tons—The Story Behind The Legend." http://www.ernieford.com/Sixteen%20Tons.htm

U.S. Army Corps of Engineers. "Tar Creek, Oklahoma, Others Cooperate to Cleanup Pollution." http://www.usace.army.mil/inet/functions/cw/hot_topics/potw.htm

U.S. Federal Register. Vol. 48, No. 175. Thursday, September 8, 1983. Rules and Regulations, 40658-40673. http://www.epa.gov/superfund/sites/npl/f830908.htm

U.S. Geological Survey. "Lead Statistics and Information." *Mineral Commodity Summaries 2005,* pp.http://minerals.usgs.gov/minerals/pubs/commodity/lead

_____. "Project Title: Assessment of Ground-Water Flow and Recharge in the Boone Aquifer in Ottawa County, Oklahoma." http://ok.water.usgs.gov/roj/boone.aquifer.html

_____. "Zinc Statistics and Information." *Mineral Commodity Summaries 2005—* 188–189.http://minerals.usgs.gov/minerals/pubs/commodity/zinc

U.S. Supreme Court. U.S. v. Charles F. Noble, John M. Cooper, A. J. Thompson, A. S. Thompson, and V.

E. Thompson, 237 U.S. 74 (1915), No. 127. http://westlaw.com/print/printstream.aspx?sv=Full&prft=HTMLE&mt=mt=Westlaw&f...

_____. George W. Choate et.al., plffs. In err., v. M. F. Trapp, Secretary of the State

Board of Equalization, et.al., 224 U.S. 665 (1912), No. 809. http://caselaw.lp.findlaw.com/scripts/printer_friendly.pl?page=us/224/665.html

Washington DC—A National Register of Historic Places Travel Itinerary. "US Capitol." http://www.cr.nps.gov/nr/travel/wash/dc76.htm

Winter, Mark. "Zinc." The University of Sheffield and WebElements, Ltd., UK. http://www.webelements.com/webelements/scholar/elements/zinc/history.html.

World Mysteries—Mystic Places. "The Great Pyramid of Giza in Egypt" http://www.world-mysteries.com/mpl_2.htm

Interviews

Hile, Louis. Interviewed Louis Hile on May 22, 2003, at the Picher Mining Museum where Mr. Hile volunteered his time. Mr. Hile spent many years in the Picher Mining Field.

END NOTES

Preface

1   James Burke, *The Pinball Effect*, (Boston: Little, Brown & Co., 1996), p. 3.

Introduction—Beginning of a Journey

2   Shaun Schafer, "Ride to offer 'Toxic Tour'," *Tulsa World*, May 1, 2003, A-1

3   Ibid., A-14

4   The Joplin Globe Online Edition, "Tar Creek—In our view," February 3, 2002, http://

www.joplinglobe.com/archives/2002/020203/headline/story4.html (accessed April 17, 2003).

5   Sierra Club, "Communities at risk: Ottowa [sic], Oklahoma," http://www.sierra-club.org/communities/Oklahoma/ (accessed April 17, 2002).

6   Shaun Schafer, "Toiling with tainted soil," Tulsa World, April 26, 2003, A-1

The Quapaw—The Downstream People

7   Genesis 13:6 (KJV)

8   W. David Baird, The Quapaw Indians—A History of the Downstream People, (Norman, Oklahoma: University of Oklahoma Press, 1980), pp. 6–8.

9   George Sabo III, "The Osage Indians," Historic Native Americans in the Mississippi Valley, Arkansas Archeological Survey, Revised July, 2001, http://www.uark.edu/depts/contact/osage.html (accessed August 12, 2006).

10 Velma Seamster Nieberding, The Quapaws (Those who went downstream), (Miami, Oklahoma: Dixons, Inc., 1976), p. 2.

11 Thomas Nuttall, *A Journal of Travels into the Arkansas Territory during the year 1819*, ed. Savoie Lottinville, (Norman, Oklahoma: University of Oklahoma Press, 1980), p. 91; Nieberding, pp. 1–2.

12 Official Quapaw Website: "Town Structure," http://geocities.com/Athens/Aegean/1388/tow.html?20072(accessed February 2, 2007); Baird, *The Quapaw Indians—A History of the Downstream People*, pp. 10–11.

13 Baird, *The Quapaw Indians—A History of the Downstream People*, pp. 10–11.

14 Morris S. Arnold, *The Rumble of a Distant Drum*, (Fayetteville, Arkansas: The University of Arkansas Press, 2000), p. 7.

15 Ibid., pp. 3–4.

16 Vern E. Thompson, *Brief History of the Quapaw Tribe of Indians*, (Pittsburg,

Kansas: Mostly Books, 1994), p. 21. Originally published 1937; Nuttall, p. 97; Baird, *The Quapaw Indians—A History of the Downstream People*, p. 12.

17 W. David Baird, *The Quapaws*, (New York: Chelsea House Publishers, 1989), pp. 8, 10.

18 Baird, *The Quapaws—A History of the Downstream People*, pp. 19–20.

19 Jean-Bernard Bossu, *New Travels in North America*, Samuel Dorris Dickinson, trans., ed., (Natchitoches, Louisiana: Northwestern State University Press, 1982), pp. v., ix.

20 Old State House Museum, "Teacher Supplement to the Arkansas News," *The Arkansas News Teachers Guide*, (Spring 1988), 2.

21 Thompson, p. 21; Nuttall, p. 97; Baird, *The Quapaw Indians—A History of the Downstream People*, p. 12.

22 George Sabo III, "Inconsistent Kin: French-Quapaw Relations at Arkansas Post," Arkansas Archeological Survey Research Series No. 40, *Arkansas Before the Americans*, Hester A. Davis, ed., (Wrightsville, Arkansas: Arkansas Department of Corrections, 1991), p. 119.

23 Ibid., p. 113.

24 Ibid., p. 118.

25 Ibid., p. 128.

26 Baird, The Quapaw Indians—A History of the Downstream People, pp. 13–15.

27 Arnold, p. 81.

28 Ibid., p. 22.

29 George E. Hyde, *Indians of the Woodlands—From Prehistoric Times to 1725*, (Norman, Oklahoma: University of Oklahoma Press, 1962), p. 49.

30 Baird, *The Quapaw Indians—A History of the Downstream People*, pp. 17–18.

31 Nieberding, *The Quapaws*, pp. 17–18.

32 Edna Kenton, ed., *The Indians of North America*, Vol. 2, (New York: Harcourt, Brace & Company, 1927), pp. 258–259.

33 Nieberding, *The Quapaws*, pp. 17–18.

34 Kenton, *The Indians of North America*, Vol. 2, pp. 280–282.

35 Baird, *The Quapaw Indians—A History of the Downstream People*, pp. 21–22.

36 Kenton, *The Indians of North America*, Vol. 2, p. 281; Nieberding, *The Quapaws*, p. 18.

37 Dan F. Morse, "On the Possible Origin of the Quapaws in Northeast Arkansas," Arkansas Archeological Survey Research Series No. 40, *Arkansas Before the Americans*, Hester A. Davis, ed., (Wrightsville, Arkansas: Arkansas Department of Corrections, 1991), p. 43.

38 Baird, *The Quapaw Indians—A History of the Downstream People*, p. 8.

39   Anna Lewis, *Along the Arkansas,* (Dallas: The Southwest Press, 1932), p. 12.

40   Baird, *The Quapaw Indians–A History of the Downstream People,* pp. 22–23.

41   Nieberding, *The Quapaws,* p. 19.

42   Baird, *The Quapaw Indians—A History of the Downstream People,* pp. 24–25; Patricia Galloway, "Couture, Tonti, and the English-Quapaw Connection, A Revision," Arkansas   Archeological Survey Research Series No. 40, *Arkansas Before the Americans,* Hester A. Davis, ed., (Wrightsville, Arkansas: Arkansas Department of Corrections, 1991), p. 75.

43   Baird, *The Quapaw Indians—A History of the Downstream People,* pp. 24–26.

44   Ibid., p. 27.

45   Ibid., pp. 28–31.

46   Nieberding, *The Quapaws,* p. 21.

47   Baird, *The Quapaw Indians—A History of the Downstream People,* pp. 27, 31, 38.

48   Richard B. Morris, ed., "The Colonies and the Empire, 1624–1775," *Encyclopedia of American History,* (New York: Harper & Brothers, 1953), pp. 65, 67.

49   Ibid., p. 70.

50   Baird, *The Quapaw Indians—A History of the Downstream People,* pp. 37–38.

51   Charles C. Mann, *1491,* (New York: Alfred A. Knopf), pp. 31–32.

52   Ibid., p. 35.

53   Hyde, p. 269.

54   C. Matthew Snipp, *American Indians: The First of This Land,* (New York: Russell Sage Foundation, 1989)pp. 23–25.

55   Arnold, pp. xxvi-xxviii.

56   Baird, *The Quapaw Indians—A History of the Downstream People,* pp. 27, 31, 38.

57   David McCullough, *John Adams,* (New York: Simon & Schuster, 2001), p. 118.

58   Ibid., pp. 136–137.

59   Ibid., p. 136.

60   Baird, *The Quapaw Indians—A History of the Downstream People,* p. 39.

61   Ibid, pp. 40–44.

62   Ibid., pp. 44–45, 47, 49.

63   Ibid., p. 50.

64   Richard B. Morris, ed., "The Early National and Ante-Bellum Periods, 1789–1860," *Encyclopedia of American History,* p. 132.

65   Joseph Harriss, "Westward Ho!," *Smithsonian,* Vol. 34, Number 1, (April 2003), 104.

66   Ibid., pp. 105–106, 108.

67   Ibid., p.102.

The Quapaw—Old Americans or New Americans?

68 Psalm 55:7 (KJV).

69 Baird, *The Quapaw Indians-A History of the Downstream People,* pp. 51–53.

70 Nieberding, *The Quapaws,* pp. 56–57.

71 Dr. David Stewart and Dr. Ray Knox, *The Earthquake That Never Went Away,* (Marble Hill, Missouri: Gutenberg-Richter Publications, 1993), pp. 17–19.

72 Ibid., pp. 20–21.

73 Ibid., p. 23.

74 Nuttall, pp.55–56.

75 Nieberding, *The Quapaws,* p. 57.

76 Baird, *The Quapaw Indians—A History of the Downstream People,* p. 53.

77 Landon Y. Jones, "Iron Will," *Smithsonian,* Vol. 33, Number 5, August 2002, 98.

78 Ibid., p. 98.

79 Ibid., pp. 103–104.

80 Ibid., pp. 96, 100.

81 Baird, *The Quapaw Indians—A History of the Downstream People,* pp. 53–55.

82 Jones, p. 96.

83 Baird, *The Quapaw Indians—A History of the Downstream People,* p. 55.

84 Ibid., pp. 55–57.

85 Charles J. Kappler, comp. and ed., *Indian Affairs: Laws and Treaties,* Vol. II, (Washington: Government Printing Office, 1904), pp. 160–161.

86 Nieberding, *The Quapaws,* P. 62.

87 Baird, *The Quapaw Indians—A History of the Downstream People,* p. 57.

88 Nieberding, *The Quapaws,* p. 62.

89 Vern E. Thompson, "A History of the Quapaw," *The Chronicles of Oklahoma,* Vol. XXXIII, (Oklahoma City, Oklahoma: Oklahoma Historical Society, 1955), p. 366.

90 Baird, *The Quapaw Indians—A History of the Downstream People,* pp. 64–65.

91 Ibid., pp. 61–62.

92 Harriss,102.

93 Nieberding, *The Quapaws,* p. 65.

94 Ibid., p. 65.

95 Thompson, "A History of the Quapaw," pp. 366–367.

96 Ibid., p. 367.

97 Ibid, p. 368.

98 Baird, The Quapaw Indians—A History of the Downstream People, p. 67.

99   Ibid., p. 65.

100 Nieberding, *The Quapaws*, p. 67.

101 Baird, *The Quapaw Indians—A History of the Downstream People*, pp. 67–68.

102 Nieberding, *The Quapaws*, p. 67.

103 Thompson, "A History of the Quapaw," p. 369.

104 Kappler, comp. and ed., *Indian Affairs: Laws and Treaties*, Vol. II, pp. 210–211.

105 Linda Parker, "Indian Colonization in Northeastern and Central Indian Territory," *The Chronicles of Oklahoma*, Vol. LIV, (Oklahoma City, Oklahoma: Oklahoma Historical Society, Spring 1976), p. 108.

106 Baird, *The Quapaw Indians—A History of the Downstream People*, pp. 68–69.

107 Ibid., pp. 70–71.

108 "The Natural Environment: The Delta and Its Resources," Nile of the New World, *Draft Heritage Study and Environmental Assessment*, Vol. 2, http://www.cr.nps.gov/delta/dhsea.htm (accessed May 19, 2006).

109 Nieberding, *The Quapaws*, p. 74.

110 Baird, *The Quapaw Indians—A History of the Downstream People*, pp. 71–72.

111 Carolyn Thomas Foreman, "Education Among the Quapaws 1829–1875," *The Chronicles of Oklahoma*, Vol. XLII, (Oklahoma City, Oklahoma: Oklahoma Historical Society, Summer 1964), p. 15.

112 Nieberding, *The Quapaws*, p. 75, 77.

113 Baird, *The Quapaw Indians—A History of the Downstream People*, pp. 74–75.

114 Ibid., pp. 75–77

115 Kappler, comp. and ed., *Indian Affairs: Laws and Treaties*, Vol. II, pp. 395–397.

116 Baird, *The Quapaw Indians—A History of the Downstream People*, pp. 77–78.

117 Nieberding, *The Quapaws*, p.84.

118 Baird, *The Quapaw Indians—A History of the Downstream People*, pp. 77–79.

119 Ibid., pp. 78–79.

120 Nieberding, *The Quapaws*, p. 89.

121 Grant Foreman, *Advancing the Frontier*, (Norman, Oklahoma, University of Oklahoma Press, 1933), pp. 165–166.

122 Grant Foreman, ed. and anno., *A Traveler in Indian Territory*, (Norman, Oklahoma, University of Oklahoma Press, 1930), p. 257.

123 Baird, *The Quapaw Indians—A History of the Downstream People*, p. 87.

124 Commissioner of Indian Affairs, *Annual Report, 1839–1840*, (Washington: J. Gideon, Jr., Printer, 1839), p. 154. (University of Tulsa, McFarlin Library, Microfilm M no. 00345).

125 Baird, *The Quapaw Indians—A History of the Downstream People*, p. 89.

126 Ibid., pp. 93–96.

127 Arrell Morgan Gibson, *Oklahoma—A History of Five Centuries*, Second Edition, (Norman, Oklahoma: University of Oklahoma Press, 1981), pp. 118–120.

128 Baird, *The Quapaw Indians—A History of the Downstream People*, p. 97.

129 Gibson, *Oklahoma—A History of Five Centuries*, p.121.

130 Baird, *The Quapaw Indians—A History of the Downstream People*, pp.97–98.

131 Nieberding, *The Quapaws*, p. 100.

132 Baird, *The Quapaw Indians—A History of the Downstream People*, pp. 102–104; Commissioner of Indian Affairs, *Report on Indian Affairs, 1865*, (Washington: Government Printing Office, 1865), pp. 294–295. (University of Tulsa, McFarlin Library, Microfilm M no. 00345).

The Quapaw—White Man's World

133 1 Chronicles 29:15 (KJV).

134 Frank H. Harris, "Neosho Agency 1838–1871," *The Chronicles of Oklahoma*, Vol. XLIII, (Oklahoma City, Oklahoma: Oklahoma Historical Society, Spring, 1965), pp. 54–55.

135 Richard B. Morris, ed., "The Civil War and Reconstruction, 1862–1877," *Encyclopedia of American History*, (New York: Harper & Brothers, 1953), p. 245.

136 Baird, *The Quapaw Indians—A History of the Downstream People*, p. 106

137 Commissioner of Indian Affairs, *Report on Indian Affairs, 1867*, p. 324.

138 Baird, *The Quapaw Indians—A History of the Downstream People*, pp. 106–107

139 Nieberding, The Quapaws, p. 104.

140 Baird, *The Quapaw Indians—A History of the Downstream People*, p. 107.

141 41st Congress, 3rd Session, House Report 39, 51.

142 Kappler, comp. and ed., *Indian Affairs: Laws and Treaties*, Vol. II, pp.961–962; Baird, *The Quapaw Indians—A History of the Downstream People*, pp. 107–108.

143 Baird, *The Quapaw Indians—A History of the Downstream People*, pp. 110–111; 41st Congress, 3rd Session, House Report 39, 51–52.

144 41st Congress, 3rd Session, House Report 39, VII-IX.

145 Commissioner of Indian Affairs, *Report on Indian Affairs, 1868*, p. 272.

146 Baird, *The Quapaw Indians—A History of the Downstream People*, pp. 108–109; Commissioner of Indian Affairs, *Report on Indian Affairs, 1869*, p. 38.

147 Commissioner of Indian Affairs, *Report on Indian Affairs, 1869*, pp. 380–381.

148 Ibid., p. 382.

149 Ibid.

150 42nd Congress, 2nd Session, House Executive Documents 276, 1–2; Baird, *The*

Quapaw Indians—A History of the Downstream People, p. 110; Commissioner of Indian Affairs, *Report on Indian Affairs, 1872*, p. 243.

151 Carolyn Thomas Foreman, "Education Among the Quapaws," *The Chronicles of Oklahoma*, Vol. XXV, (Oklahoma City, Oklahoma: Oklahoma Historical Society, Spring 1947), p. 16.

152 Baird, *The Quapaw Indians—A History of the Downstream People*, pp. 89–90.

153 Foreman, "Education Among the Quapaws," *The Chronicles of Oklahoma*, Vol. XXV, pp. 16–17; Baird,

154 Foreman, "Education Among the Quapaws," p. 22.

155 Ibid., pp. 21–22.

156 Baird, *The Quapaw Indians—A History of the Downstream People*, p. 93.

157 Foreman, "Education Among the Quapaws," *The Chronicles of Oklahoma*, Vol. XXV, p. 22, 25.

158 Baird, *The Quapaw Indians—A History of the Downstream People*, p. 112.

159 Joe C. Jackson, "Schools Among the Minor Tribes in Indian Territory," *The Chronicles of Oklahoma*, Vol. XXXII, (Oklahoma City, Oklahoma: Oklahoma Historical Society, 1954), p. 58.

160 Baird, *The Quapaw Indians—A History of the Downstream People*, pp. 112–113; Commissioner of Indian Affairs, *Report on Indian Affairs, 1872*, p. 243; Commissioner of Indian Affairs, *Report on Indian Affairs, 1873*, p. 214; Commissioner of Indian Affairs, *Report on Indian Affairs, 1874*, p. 228.

161 Jackson, "Schools Among the Minor Tribes in Indian Territory," *The Chronicles of Oklahoma*, Vol. XXXII, p. 64.

162 Baird, *The Quapaw Indians—A History of the Downstream People*, pp. 114, 116; Commissioner of Indian Affairs, *Report on Indian Affairs, 1873*, pp. 213–214; Commissioner of Indian Affairs, *Report on Indian Affairs, 1874*, p. 228; Commissioner of Indian Affairs, *Report on Indian Affairs, 1876*, p. 57.

163 Commissioner of Indian Affairs, *Report on Indian Affairs, 1872*, p 243.

164 Commissioner of Indian Affairs, *Report on Indian Affairs, 1873*, p. 214.

165 Commissioner of Indian Affairs, *Report on Indian Affairs, 1874*, p. 228.

166 Commissioner of Indian Affairs, *Report on Indian Affairs, 1875*, p. 281.

167 Commissioner of Indian Affairs, *Report on Indian Affairs, 1876*, p. 57.

168 Baird, *The Quapaw Indians—A History of the Downstream People*, pp. 117–118; Commissioner of Indian Affairs, *Report on Indian Affairs, 1877*, p. 103.

169 Carolyn Thomas Foreman, "Lewis Francis Hadley: The Long-Haired Sign Talker," *The Chronicles of Oklahoma*, Vol. XXVII, (Oklahoma City, Oklahoma: Oklahoma Historical Society, Spring 1949), pp. 42, 47.

170 Ibid., 43.

171 Commissioner of Indian Affairs, *Report on Indian Affairs, 1877*, p. 103.

172 Commissioner of Indian Affairs, *Report on Indian Affairs, 1878*, p. 65.

173 Baird, *The Quapaw Indians—A History of the Downstream People*, p. 117; Commissioner of Indian Affairs, *Report on Indian Affairs, 1877*, p. 103.

174 Nieberding, *The Quapaws*, p. 110.

175 Baird, *The Quapaw Indians—A History of the Downstream People*, p. 119.

176 Stan Hoig, *The Oklahoma Land Rush of 1889*, (Oklahoma City, Oklahoma: Oklahoma Historical Society, 1984), p. 14.

177 Commissioner of Indian Affairs, *Report on Indian Affairs, 1879*, pp. 75–76, 78.

178 Baird, *The Quapaw Indians—A History of the Downstream People*, pp. 120–121; Commissioner of Indian Affairs, *Report on Indian Affairs, 1879*, p. 78.

179 Commissioner of Indian Affairs, *Report on Indian Affairs, 1868*, p. 272.

180 Baird, *The Quapaw Indians—A History of the Downstream People*, pp. 123–125; Commissioner of Indian Affairs, *Report on Indian Affairs, 1868*, p. 272.

181 Nieberding, *The Quapaws*, p. 121.

182 Commissioner of Indian Affairs, *Report on Indian Affairs, 1879*, p. 75.

183 Commissioner of Indian Affairs, *Report on Indian Affairs, 1880*, p. 86.

184 Baird, *The Quapaw Indians—A History of the Downstream People*, pp. 127–128.

185 Ibid., p. 130.

186 Ibid., pp. 127, 130–131, 133.

187 Ibid, pp. 133–135.

188 Ibid., p. 133.

189 Nieberding, *The Quapaws*, pp. 125–126.

190 Commissioner of Indian Affairs, *Report on Indian Affairs, 1889*, p. 198.

191 51st Congress, 1st Session, House Report 559, 1–2.

192 Baird, *The Quapaw Indians—A History of the Downstream People*, pp. 135–136.

193 Ibid., pp. 137–138.

194 Henry Steele Commager, ed., *Documents of American History*, Vol. II since 1865, (New York, F. S. Crofts & Co., 1935), pp. 124–125.

195 Baird, *The Quapaw Indians—A History of the Downstream People*, pp. 138–140; Testimony of Abner W. Abrams, Committee on Indian Affairs, U. S. Senate, March 5, 1892, 17; 52nd Congress, 1st Session, House Report 2040, 1–2; 52nd Congress, 1st Session, Senate Report 615, 1–3.

196 Baird, *The Quapaw Indians—A History of the Downstream People*, pp. 140–142;

197 Ibid., pp. 145–147; 53rd Congress, 3rd Session, *Congressional Record*, 1233.

198 Nieberding, *The Quapaws*, p. 128.

199 Baird, *The Quapaw Indians—A History of the Downstream People*, pp. 146–147.

200 Nieberding, *The Quapaws*, p. 135.

Indian Country to Statehood—Life on the Edge

201 Joshua 23:12–13 (KJV).

202 Michael Wallis, *Oil Man*, (New York: Doubleday, 1988), pp. 55–57.

203 Morris, p. 117.

204 Henry Steele Commager, ed., *Documents of American History*, Vol. 1 to 1865, (New York: F. S. Crofts & Co., 1935), P. 132.

205 H. Wayne Morgan and Anne Hodges Morgan, *Oklahoma—A History*, (New York: W. W. Norton & Company, 1984), p. 4.

206 Rennard Strickland, *The Indians of Oklahoma*, (Norman, Oklahoma: University of Oklahoma Press, 1980), pp. 2–3.

207 *Encyclopedia of North American Indians—Indian Territory*, http://college.hmco. com/history/ readerscomp/naind/html/na_016600_indianterrit.htm (accessed September 27, 2003).

208 Morgan, pp. 46–47.

209 Lloyd H. McGuire, Jr., *Birth of Guthrie*, (San Diego, California: Crest Offset Printing Company, 1998), p. 10.

210 Morgan, *Oklahoma—A History*, pp. 24–25.

211 Strickland, p. 19.

212 McGuire, pp. 11–13.

213 Hoig, p. 4.

214 Ibid., pp. 4–7.

215 Ibid., pp. 7–12

216 Ibid., pp. 14–16.

217 McGuire, p. 17.

218 Duane Champagne, ed., *The Native North American Almanac*, (Detroit, Michigan: Gale Research, Inc. 1994), p. 213.

219 Hoig, pp. 61–66.

220 Morgan, *Oklahoma—A History*, pp. 49–50.

221 McGuire, p. 21.

222 Joseph B. Thoburn and Muriel H. Wright, Oklahoma—A History of the State and Its People, Volume II, (New York: Lewis Historical Publishing Company, Inc., 1929), pp. 543–546; McGuire, pp. 21–23.

223 Odie B. Faulk, *Oklahoma—Land of the Fair God*, (Northridge, California: Windsor Publications, Inc., 1986), p. 126.

224 McGuire, p. 47.

225 Morgan, *Oklahoma—A History*, p. 53.

226 Hoig. pp. 193–194.

227 Ibid., pp. 195–197.

228 Morgan, *Oklahoma—A History*, pp. 64–66.

229 *Encyclopedia of North American Indians—Indian Territory.*

230 Morgan, *Oklahoma—A History*, p. 58.

231 Faulk, p. 130.

232 D. Earl Newsom, *The Cherokee Strip—Its History & Grand Opening*, (Stillwater, Oklahoma: New Forums Press, Inc., 1992) p. 1.

233 Ibid., pp. 40–41.

234 D. E. Quick, "A Great Race," *The Kendall Collegian*, (May 1906), 93–94. McFarlin Library College Files, University of Tulsa, Tulsa Oklahoma.

235 Newsom, p. 42

236 Quick, pp.93–94

237 Ibid., pp. 94–95.

238 Newsom, p. 46.

239 Faulk, pp. 131–132.

240 Ibid., p. 132.

241 Anne Hodges Morgan and H. Wayne Morgan, Ed., *Oklahoma—New Views of the Forty-Sixth State*, (Norman, Oklahoma: University of Oklahoma Press, 1982), pp. 35–36.

242 Faulk, pp. 126–127.

243 Ibid., pp. 131–132.

244 *Encyclopedia of North American Indians—Indian Territory.*

245 Strictland, pp. 48–49.

246 Faulk, pp. 149–150.

247 Ibid., pp. 150–153.

248 Ibid., pp. 153–156.

249 Strictland, pp. 50–53.

250 Faulk, pp. 156–157.

251 Edmund Morris, *Theodore Rex*, (New York: The Modern Library, 2001), p. 539.

252 Morgan, *Oklahoma—A History*, pp. 90–91.

253 Faulk, p. 157.

254 Edward Everett Dale, "Two Mississippi Valley Frontiers," *Chronicles of Oklahoma*, Vol. XXVI, (Oklahoma City, Oklahoma: Oklahoma Historical Society, Winter 1948–49), p. 382.

Wildcatters and Prospectors

255 Ezekiel 34:27 (KJV).

256 Wallis, p. 66.

257 Ibid., pp. 58–60.

258 Ibid., pp. 63–66.

259 Ibid., pp. 47–48.

260 Ibid., p. 67.

261 James Tobin, "To Fly!," *Smithsonian*, Vol. 34, Number 1, April 2003, 61–62.

262 Wallis, pp. 468–469.

263 Ibid., pp. 58–59.

264 Ibid., p. 146.

265 Faulk, p. 132.

266 Champagne, p. 214.

267 Gibson, *Oklahoma—A History of Five Centuries*, p. 270.

268 Wallis, p. 152.

269 Ibid., pp. 314–315.

270 Baird, *The Quapaw Indians—A History of the Downstream People*, pp. 150–151.

271 Nieberding, *The Quapaws*, p. 132.

272 Baird, *The Quapaw Indians—A History of the Downstream People*, p. 150.

273 Arrell M. Gibson, *Wilderness Bonanza*, (Norman, Oklahoma: University of Oklahoma Press, 1972), p. 15.

274 Ibid, pp. 18–19.

275 Burke, pp. 9–10; Morris, *Encyclopedia of American History*, p. 207.

276 Velma Nieberding, *The History of Ottawa County*, (Marceline, Missouri: Walsworth Pub. Co., 1983), p. 57.

277 Dolph Shaner, *The Story of Joplin*, (New York: Stratford House, Inc., 1948), pp. 3–4.

278 Gibson, *Wilderness Bonanza*, pp. 19–20.

279 Shaner, pp. 1–4.

280 Ibid., p. 2.

281 Gibson, *Wilderness Bonanza*, pp. 19–21.

282 Howard W. Blosser, *Prairie Jackpot*, (Webb City, Missouri, 1973), pp.13.

283 Gibson, *Wilderness Bonanza*, pp. 25–26.

284 Ibid., pp. 31–32.

285 Ibid., pp. 34–36.

286 Ibid., p. 37.

287 Galena Bicentennial Committee, *Pioneer Days of Galena*, (Galena, Kansas: Tri-State Printing Co., 1984), pp.1–2, 5–7.

288 Ibid., p. 7.

289 Ibid., p. 11.

290 Ben Moody, "O. W. (Ol) Sparks, a mining man," *Tri-State Tribune*, August 9, 1990, 12.

291 Ibid.

292 Blosser, pp.1–3.

293 Nieberding, *The History of Ottawa County*, pp.59–60.

294 Ibid., pp. 61–61.

295 Gibson, *Wilderness Bonanza*, p. 39.

296 Samuel Weidman, *Miami-Picher Zinc-Lead District*, (Norman, Oklahoma: University of Oklahoma Press, 1932), p. 45.

297 Nieberding, *The History of Ottawa County*, p. 64.

298 Weidman, pp.46–47.

299 Nieberding, *The History of Ottawa County*, pp. 68–71.

300 Ibid., pp. 71–72.

Boomtown!

301 1 Timothy 6:9 (KJV)

302 Betty Pulley, "The History of Picher," *The Chat Pile*, June 10, 1998, 14.

303 Blosser, pp. 15–17.

304 Ibid., p. 38.

305 Ibid., pp. 26–31.

306 Ibid., p. 20.

307 Ibid., pp. 43–48.

308 Ibid., pp. 44–51.

309 Ibid., pp. 63, 66.

310 Pulley, p. 14.

311 Gibson, *Wilderness Bonanza*, p. 40.

312 Ibid.

313 Nieberding, *The History of Ottawa County*, p. 81.

314 Weidman, fold out chart between pp. 47–48.

315 Nieberding, *The History of Ottawa County*, p. 81.

316 Ibid., p. 32.

317 Pulley, p. 14.

318 Ibid.

319 "City of Picher born out of a happy accident," *Tri-State Tribune*, May 30, 2002, p. 2b.

320 Pulley, p. 14.

321 Ibid.

322 Allan Matthews, "Picher growing pains—the first dozen years," *Tri-State tribune*, August 9, 1990, 14-B.

323 Ibid.

324 Pulley, p. 14.

325 Matthews, p. 14-B

326 Ibid.

327 *World Almanac and Book of Facts 1999*, (Mahwah, New Jersey: World Almanac Books, 1998), p. 582.

328 Spencer C. Tucker, *The Great War 1914–18*, (Bloomington & Indianapolis: Indiana University Press, 1998), pp. 11–12.

329 Ibid., pp. 15–16.

330 Ibid., p. 13.

331 Ian V. Hogg, *Allied Artillery of World War One*, (Wiltshire, England: The Crowood Press, 1998), p. 17.

332 Guy Hartcup, *The War of Invention*, (London: Brassey's Defence Publishers, 1988), p. 44.

333 Hogg., p. 17.

334 Ibid., pp. 7–8.

335 Ibid., pp. 203–205.

336 Tucker, p. 13.

337 John Ellis, "The War in the Trenches: Guns and Gas," *World War I*, Donald J. Murphy, ed., (San Diego, California: Greenhaven Press, Inc., 2002), p. 111

338 Tucker, p. 13.

339 Ibid., p. 176.

340 Donald J. Murphy, ed. *World War I*, (San Diego, California: Greenhaven Press, Inc., 2002), pp. 272-273.

341 Marie Rentfrow, "The Town that Jack Built," *The Chat Pile*, June 10, 1998, 28.

342 Gibson, *Wilderness Bonanza*, p. x.

343 Tonnage refers to short tons. A short ton is 2,000 pounds as opposed to the long ton of 2,240 pounds.

344 Weidman, fold out chart between pp. 47–48.

345 Ibid., p. 48.

Brother Lead and Cousin Zinc

346 Genesis 1:9 (KJV)

347 "City of Picher born out of a happy accident," *Tri-State Tribune*, May 30, 2002, 4b.

348   Susan Watt, *Lead,* (New York, New York: Benchmark Books, 2002), p. 8–9

349   Ibid., p. 5.

350   Ibid., p. 9.

351   Ibid., pp. 10–11.

352   C. J. S. Thompson, *The Lure and Romance of Alchemy,* (New York: Bell Publishing Company, 1990), pp. 14–15.

353   Watt, p. 7.

354   Weidman, p. 77.

355   Gibson, *Wilderness Bonanza,* pp. 5–6.

356   Ibid., pp. 4–5.

357   Weidman, p. 69.

358   Gibson, *Wilderness Bonanza,* pp.10–13.

359   Ibid., pp. 8–9.

360   Watt, p. 4.

361   Weidman, p. 54.

362   Watt, pp. 13, 27.

363   Ibid., pp. 25–26

364   Ibid., p. 26.

365   U.S. Geological Survey, "Lead Statistics and Information," *Mineral Commodity Summaries 2005,* pp. 94–95 http://minerals.usgs.gov/minerals/pubs/commodity/lead (accessed March 17, 2005).

366   Ibid.

367   Alan W. Cramb, "A Short History of Metals," Carnegie Mellon University, http://neon.mems.cmu/ cramb/processing/history.html (accessed April 2, 2005).

368   Genesis 4:22 (KJV)

369   Mark Winter, "Zinc," The University of Sheffield and WebElements, Ltd., UK, http://www.webelements.com/webelements/scholar/elements/zinc/history.html. (accessed April 2, 2005).

370   Jefferson Lab, "It's Elemental.," http://education.jlab.org/itselemental/ele030.html (accessed April 2, 2005).

371   Brainy Encyclopedia, "Zinc," http://www.brainyencyclopedia.com/encyclopedia/z/zi/sinc.html (accessed April 2, 2005).

372   Los Angeles Chinese Learning Center, "General Information of Zinc," http://chinese- school.netfirms.com/Zinc-information.html (accessed April 2, 2005).

373   Jefferson Lab, http://education.jlab.org/itselemental/ele030.html (accessed April 2, 2005).

374   Mark Winter, "Zinc," The University of Sheffield and WebElements, Ltd, UK,

http://www.webelements.com/webelements/scholar/elements/zinc/history. html (accessed April 2, 2005).

375 "City of Picher born out of happy accident," *Tri-State Tribune*, May 30, 2002, 2b.

376 Los Angeles Chinese Learning Center, http://chinese-school.netfirms.com/ Zinc-information.html (accessed April 2, 2005).

377 Marie Rentfrow, "The Town that Jack Built, Tri-State Tribune, June 10, 1998, 28.

378 Los Angeles Chinese Learning Center, http://chinese-school.netfirms.com/ Zinc-information.html (accessed April 2, 2005).

379 Gibson, *Wilderness Bonanza*, p. 8.

380 Smithsonian Institution, "James Smithson—Founder of the Smithsonian Institution," http:// www.si.edu/archives/documents/smithson.htm (accessed April 19, 2005).

381 Nina Burleigh, *The Stranger and the Statesman*, (New York: William Morrow, 2003), p. 182–183.

382 Ibid., p. 250.

383 Thomas G. and Virginia L. Aylesworth, *Washington—The Nation's Capitol*, (New York: Smithmark Publishers, Inc., 1991), p. 28.

384 U.S. Geological Survey, "Zinc Statistics and Information," *Mineral Commodity Summaries 2005*, pp. 188–189, http://minerals.usgs.gov/minerals/pubs/commodity/zinc (accessed April 20, 2005).

Into the Pits

385 Psalm 143:7 (KJV)

386 Weidman, p. 92.

387 Gibson, Wilderness Bonanza, pp. 67, 79, 91.

388 Ibid., p.68.

389 Ibid., p. 42.

390 Ibid., p. 42–44.

391 Weidman, p. 96.

392 Ibid.

393 Gibson, Wilderness Bonanza, p. 77.

394 Weidman, pp. 97–99.

395 Frank Wood, "Language of the mines," Tri-State Tribune, June 12, 2003, 2.

396 Ibid.

397 Gibson, Wilderness Bonanza, p. 89.

398 Ibid.

399 Ibid.

400 Charles Chestnut, "Those days in the mines," Tri-State Tribune, August 9, 1990, 1-B

401 Weidman, p. 101.

402 Interview of Louis Hile on May 22, 2003 at the Picher Mining Museum where Mr. Hile volunteered his time.

403 Weidman, p. 101.

404 Ibid.

405 Nieberding, "Mules," The History of Ottawa County, p. 95.

406 Amerillis Brookshire, "Tobacco chewing mule," Tri-State Tribune, June 12, 2003, 2.

407 "The Story of Old Toby, the last of the lead mine mules," Tri-State Tribune, August 9, 1990, 1.

408 Frank D. Wood, "Moving Mules," Tri-State Tribune, August 9, 1990, 18-B.

409 Gibson, Wilderness Bonanza, p. 89.

410 Weidman, p. 102.

411 Gibson, Wilderness Bonanza, p. 90.

412 Ibid.

413 Wood, "Languages of the mine," Tri-State Tribune, 2.

414 TEF Enterprises, "Sixteen Tons—The Story Behind The Legend," http://www.ernieford.com/ Sixteen%20Tons.htm (accessed August 23, 2005).

Mountains of Chaff

415 Isaiah 33:11 (KJV)

416 Weidman, pp. 102–104.

417 Frank Wood, "Language of the mines," Tri-State Tribune, 2.

418 Gibson, Wilderness Bonanza, p. 100.

419 Ibid.

420 Ibid., pp. 100–101.

421 Weidman, pp. 104–105.

422 Ibid., p. 115

423 Ibid., p. 118.

424 Gibson, Wilderness Bonanza, p. 102.

425 Weidman, p. 118.

426 Gibson, Wilderness Bonanza, p. 108.

427 Ibid., p. 109.

428 Weidman, pp. 111–113

429 Gibson, Wilderness Bonanza, pp. 109–110.

430 Ibid., p. 111.

431 Weidman, pp. 50–51.

432 Gibson, Wilderness Bonanza, p. 112.

433 Ben Moody, "Yester-year In The Picher Mining Field—Building a Central Mill," Tri-State Tribune, September 8, 1988.

434 Ben Moody, "Yester-year In The Picher Mining Field—The Building of a Central Mill," Tri-State *Tribune,* September 15, 1988.

435 Ben Moody, "Yester-year In The Picher Mining Field—The Building of The Central Mill," Tri-State *Tribune,* September 22, 1988.

436 Ben Moody, "Yester-year In The Picher Mining Field—The Building of The Central Mill," Tri-State *Tribune,* September 29, 1988.

437 Ben Moody, "Yester-year In The Picher Mining Field—The Building of The Central Mill (Continued)," *Tri-State Tribune,* October 6, 1988.

438 Julie DelCour, "The town that Jack built," Tulsa World, June 22, 2003, G-1.

439 Interview of Louis Hile on May 22, 2003 at the Picher Mining Museum where Mr. Hile volunteered his time. Mr. Hile and his father worked in and around the Picher mining camp for may years.

440 Gibson, Wilderness Bonanza, p. 95.

441 Hile interview, May 22, 2003.

442 "US Capitol," Washington DC—A National Register of Historic Places Travel Itinerary, http://www.cr.nps.gov/nr/travel/wash/dc76.htm (accessed November 28, 2006).

443 "Tar Creek," The Joplin Globe Online Edition, February 23, 2002, http://joplinglobe.com/archives/2002/020203/headline/Story4.html (accessed April 17, 2003).

444 DelCour, "The town that Jack built," Tulsa World, G-1

445 Ben Moody, "Yester-year In The Picher Mining Field—Central Mill Chat Loading," Tri-State Tribune, June 30, 1988.

446 Nieberding, The History of Ottawa County, p. 73.

447 Ibid.

448 Hile interview, May 22, 2003.

449 World Mysteries—Mystic Places, "The Great Pyramid of Giza in Egypt," http://www.world-mysteries.com/mpl_2.htm (accessed August 30, 2005).

450 DelCour, "The town that Jack built," Tulsa World, G-1.

451 Kevin Jackson and Jonathan Stamp, Building the Great Pyramid, (London: Firefly Books, 2002), pp. 32–33.

452 Gibson, Wilderness Bonanza, p. 199.

Hell's Fringe

453 Ecclesiastes 2:17 (KJV)

454 Nieberding, History of Ottawa County, p. 199.

455 Ibid., p. 200.

456 John Toland, Dillinger Days, (New York: Random House, 1963; New York: Da Capo Press, Inc., 1995), pp. 249–251. Citation is to the Da Capo edition.

457 Henry Methvin—April 8, 1912-April 19, 1948, http://www.tmethvin.com/henry/ (accessed November 8, 2006)

458 Nieberding, History of Ottawa County, pp. 200–201.

459 Ibid., p. 204.

460 Ibid., p. 203

461 Bryan Burrough, Public Enemies, (New York: The Penguin Press, 2004), pp. 32.33.

462 Nieberding, History of Ottawa County, p. 203.

463 Ibid., pp. 204–205.

464 Ibid., p. 75.

465 "Joe Allen killed-fell into shaft," Miami Record Herald, March 27, 1914,1.

466 "Second Picher mine accident victim dies," Joplin News Herald, April 30, 1926.

467 "Man hurt in accident-dies here," Miami Record Herald, July 21, 1927.

468 "Ten hurt, one fatally in mine gas explosion," Joplin Globe, November 28, 1925, 1.

469 "Miner seriously hurt by falling boulder," Miami News Record, March 19, 1925, 3.

470 "Miner's fall down shaft fatal to two," Joplin News Herald, December 9, 1925, 1.

471 "Four miners plunge to their deaths in a mine shaft," Joplin Globe, October 5, 1928, 1.

472 "1 dead, 3 hurt in mine explosion," Joplin Globe, August 5, 1928, 1.

473 "Inquest tonight into death of John Hostetter," Joplin News Herald, August 25, 1926, 5.

474 "Accident Fatal to Rufus Nolan," Joplin Globe, April 28, 1925, 1.

475 "Wilbur Kell," Miami News Record, June 7, 1925.

476 "Discovery of man's body in mine shaft leads to detention of his uncle," Miami News Record, January 18, 1925, 1.

477 Iva O. Simpson, "Last Fatal Accident," Velma Nieberding, History of Ottawa County, (Marceline, Missouri: Walsworth Pub. Co., 1983), p. 93.

478 Nieberding, History of Ottawa County, p. 93.

479 Ibid., p. 96.

480 Galena Mining and Historical Museum, Galena, Kansas. Information retrieved and supplied by Carla Taylor Jordan.

481 Marjorie Ann Patton, "Women in the Mines," The Chat Pile, June 10, 1998, 15.

482 Ibid.

483 Nieberding, History of Ottawa County, p. 93.

484 "City of Picher born out of happy accident," Tri-State Tribune, May 30, 2002, 4b.

485 Nieberding, History of Ottawa County, pp. 97–98.

486 Frank Wood, "Kangaroo courts common in mines," Tri-State Tribune, May 30, 2002, 2b.

487 Wallis, pp. 180–183; "Bartlesville Sports History," http://myweb.cableone.net/gmeador/sports.htm (accessed September 7, 2002).

488 Wallis, p. 180.

489 Ben Moody, "Yester-year in the Picher mining field—Baseball in the Picher Mining Field," Tri-State *Tribune*, October 29, 1992, 4.

490 Ibid.

491 Ben Moody, "Yester-year in the Picher mining field—Mickey Mantle," Tri-State Tribune, November 5, 1992, 4.

492 Merlyn, Mickey Jr., David, and Dan Mantle with Mickey Herskowitz, A Hero All His Life, (New York: Harper Collins Publishers, 1996), pp. 9–11.

493 Ibid., pp. 48–49.

494 Mickey Mantle and Herb Gluck, The Mick, (Garden City, New York: Doubleday & Company, Inc., 1985), pp. 44–45.

495 Mantle, Merlyn, A Hero All His Life, pp. 50–55.

496 Ibid., p. 58.

497 Ibid., p. 12.

498 Ibid., p. 8.

499 James L. Lambert, "Mickey Mantle's 11th Hour Miracle," Agape Press, August 12, 2005, http:// headlines.agapepress.org/archive/8/122005f.asp (accessed August 12, 2005).

Miners' Health and Labor Strife

500 Ecclesiastes 2:11 (KJV)

501 Gibson, Wilderness Bonanza, p. 199.

502 Nieberding, History of Ottawa County, p. 76.

503 Ben Moody, "Yester-year In The Picher Mining Field—The Miners' Hard Hat," Tri-State Tribune, February 9, 1989.

504 Ben Moody, "Yester-year In The Picher Mining Field—Silicosis," Tri-State Tribune, January 26, 1989.

505 Charles B. Clayman, MD, ed., "Pneumoconiosis," Home Medical Encyclopedia, Vol. Two—I-Z, (New York: Random House, 1989), pp. 802–803.

506 "Joplin—Dec.2," Engineering and Mining Journal, Vol. XCVIII, December 5, 1914, 1062.

507 Gibson, Wilderness Bonanza, pp. 186–187.

508 R. R. Sayers, F. V. Meriwether, A. J. Lanza, and W. W. Adams, "Silicosis and Tuberculosis Among Miners of the Tri-State District of Oklahoma, Kansas, and Missouri-I," *U.S. Department of Commerce Technical Paper 545*, (Washington, D. C., Government Printing Office, 1933), pp. 4–5.

509 Ibid., p. 27.

510 Gibson, Wilderness Bonanza, pp. 194–195.

511 "Mine Town Nurse," Daily Oklahoman, November 21, 1948, 6-D.

512 Ibid.

513 Neiberding, History of Ottawa County, p. 90.

514 Ibid.

515 Frederick Lynn Ryan, Problems of the Oklahoma Labor Market, with Special Reference to *Unemployment Compensation*, (Oklahoma City, Oklahoma: Semco Color Press, Inc., 1937), pp. 11, 38.

516 Gibson, Wilderness Bonanza, p. 196.

517 George G. Suggs, Jr., Union Busting in the Tri-State, (Norman, Oklahoma: University of Oklahoma Press, 1986), p. 18.

518 Ibid., pp. 10–11.

519 U.S. Department of Commerce, Bureau of Census, Fifteenth Census of the *United States: 1930, Volume III: Part 1, Alabama-Missouri,* (Washington, D. C.: Government Printing Office, 1932), pp. 830, 1332, 1334; Volume III, Part 2, *Montana-Wyoming,* p. 554.

520 Malcolm H. Ross, Death of a Yale Man, (New York: Farrar & Rinehart, Inc, 1939), p. 185.

521 Arrell Morgan Gibson, "Poor Man's Camp," The Chronicles of Oklahoma, Vol. LX, (Oklahoma City, Oklahoma: Oklahoma Historical Society, Spring 1982), pp. 10–14.

522 Suggs, pp. 8–9.

523 Gibson, "Poor Man's Camp," The Chronicles of Oklahoma, Vol. LX, p. 15.

524 Suggs, pp. 7–8.

525 M. D. Harbaugh, "Labor Relations in the Tri-State Mining District," Mining Congress Journal, 22, June 1936, 20–21.

526 Suggs, p. 23.

Pickets, Pick Handles, and Police

527 Proverbs 4:17 (KJV)

528 Ross, pp. 189–190.

529 "Tri-State Strike Broken," News of the Industry, Engineering and Mining Journal, Vol. CXXXVI (July 1935), 347.

530 Ross, p. 190.

531 Suggs, p. 37.

532 Ibid., pp. 38–39.

533 Ibid., pp. 39–40.

534 Harbaugh, "Labor Relations in the Tri-State Mining District," 19, 24.

535 Suggs, pp. 39–41.

536 "Eagle-Picher Mining & Smelting Company, et al., Case No. C-73," Decisions and Orders of the *National Labor Relations Board*, Vol. 16, October 16–31, 1939, (Washington: United States Government Printing Office, 1940), p. 741.

537 Ross, p. 191.

538 Suggs, pp. 46–47.

539 Ibid., p. 47.

540 Ibid., pp. 54–57.

541 "3000 at union miners' meeting," Joplin Globe, May 22, 1935, 2.

542 Suggs, p. 58.

543 Ibid., p. 59.

544 "Miners to hold meeting today," Miami (Oklahoma) Daily News Record, May 26, 1935, 1.

545 "Eagle-Picher Mining & Smelting Company, et al., Case No. C-73," Decisions and Orders of the *National Labor Relations Board*, Vol. 16, p. 750.

546 Suggs, p. 62.

547 Ross, p. 193

548 Suggs, p. 74–75.

549 "Militia reaches mine field after day of rioting," Joplin Globe, May 28, 1935, 4.

550 Suggs, pp. 75, 77.

551 Ibid., p. 75–76

552 "Eagle-Picher Mining & Smelting Company, et al., Case No. C-73," Decisions and Orders of the *National Labor Relations Board*, Vol. 16, p. 744.

553 Ibid.

554 Ibid.; Militia reaches mine field after day of rioting," Joplin Globe, May 27, 1935, 4.

555 "Governor sends troops here to quell disorder," Miami (Oklahoma) Daily News Record, May 27, 1935, 1.

556 Ross, p. 195.

557 Ibid., p. 196.

558 Suggs, pp. 175–176; A. L. A. Schechter Poultry Corporation et al. v. Unites States, 295 U.S. 495 (1935). http://web2.westlaw.com/print/printstream. aspx?sv=Full&prft=HTMLE&mt=Westlaw&f…(Accessed December 11, 2006).

559 Suggs, pp. 175–176.

560 Ross, p. 203; Suggs, p. 176.

561 Ross, pp. 207–208.

562 Suggs, pp. 178–179.

563 Gibson, Wilderness Bonanza, p. 239.

564 "N. L. R. B. v. Jones & Laughlin Steel Corp., 301 U.S. 1, April 12, 1937, National Labor Relations Board, *Court Decisions Relating to the National Labor Relations Act*, Vol. I, Cases decided before December 31, 1939, (Washington, D. C.: Government Printing Office, 1944), pp. 333–334.

565 "Eagle-Picher Mining & Smelting Company, et al., Case No. C-73," Decisions and Orders of the *National Labor Relations Board*, Vol. 16, p. 758.

566 Suggs, pp. 140–143.

567 "Eagle-Picher Mining & Smelting Company, et al., Case No. C-73," Decisions and Orders of the *National Labor Relations Board*, Vol. 16, p. 728.

568 Suggs, p. 193.

569 Ibid., pp. 195–197.

570 "American Plague Spot," The New Republic, Vol. CII, No. 1, (January 1, 1940), 7–8.

571 Suggs, p. 73.

572 "Baxter tailing mill operator injured in attack by two men," Joplin Globe, May 21, 1935, 1.

573 Suggs, pp. 108, 111.

574 Ibid., pp. 112–113.

575 "Eagle-Picher Mining & Smelting Company, et al., Case No. C-73," Decisions and Orders of the *National Labor Relations Board*, Vol. 16, pp. 762–763.

576 Suggs, pp. 84–85.

577 "230 guardsmen arrive to put lid on violence in Kansas mining field," Miami (Oklahoma) Daily News-*Record*, June 9, 1935, 1.

578 "Miner is wounded," Miami (Oklahoma) Daily News-Record, June 9, 1935, 1.

579 Suggs, pp. 95–97.

580 "Tri-State Strike Broken," Engineering and Mining Journal, 347.

581 Suggs, pp. 163–164.

582 Ibid., pp. 165–167.

583 "Eagle-Picher Mining & Smelting Company, et al., Case No. C-73," Decisions and Orders of the *National Labor Relations Board*, Vol. 16, p. 763.

584 Suggs, pp. 168–169.

585 Ibid., pp. 170–171.

586 Ibid., p. 198.

587 US Department of Labor, Bureau of Labor Statistics, Wage in the Nonferrous-Metals Industry, June *1943*, (Washington, D.C.: Government Printing Office, June 1943), p. 7.

"The filthiest town know this side of Hell"

588 "The filthiest town know this side of Hell"Nieberding, *History of Ottawa County*, p. 75. Line quoted from poem written by unnamed miner about conditions in 1918 Picher.

589 Matthew 23:38 (KJV)

590 Nieberding, History of Ottawa County, p. 86–88.

591 "Picher Area Shocked by E-P News: Other firms to continue in the field," Miami (Oklahoma) Daily *News-Record*, April 7, 1957, 1.

592 "E-P suspending mine, mill operations across district," Miami (Oklahoma) Daily News-Record, April 23, 1957, 1.

593 Nieberding, History of Ottawa County, pp. 88–89.

594 "Pollution solution," Miami (Oklahoma) Daily News-Record, September 7, 1981, 2.

595 Jonna Polk, "Tar Creek is nation's largest Superfund site," U. S. Army Corps of *Engineers, Tulsa District, May 2005 Engineer Update*, http://www.hq.usace.mil/ccpa/pubs/ mayo5/STORY11.htm (accessed January 11, 2006).

596 Katherine N. Probst, Don Fullerton, Robert E. Litan, and Paul R. Portney, Footing the *Bill for Superfund Cleanups*, (Washington, D.C.: The Brookings Institution and Resources for the Future, 1995), p. 12.

597 U.S. Federal Register, Vol. 48, No. 175, Thursday, September 8, 1983, Rules and Regulations, pp. 40658- 40673, http://www.epa.gov/superfund/sites/npl/f830908.htm (accessed January 11, 2006).

598 Ibid.

599 U.S. Army Corps of Engineers, "Tar Creek, Oklahoma, Others Cooperate to Clean-up Pollution," http://www.usace.army.mil/inet/functions/cw/hot_topics/potw.htm (accessed January 11, 2006).

600 Jonna Polk, "Tar Creek is nation's largest Superfund site."

601 Katherine N. Probst, "Superfund at 25: What Remains to be Done," Resources,

Fall 2005, http://www.rff.org/rff/News/Features/Superfund-at-25.cfm (accessed January 11, 2006).

602 Ibid.

603 Ed Keheley and Mary Ann Pritchard, Report to Governor Keating's Tar Creek *Superfund Task Force by the Subsidence Committee,* July 21, 2000, p. 13.

604 "Tar Creek (Ottawa County) Oklahoma, EPA ID# OKD980629844," EPA Region 6, *Congressional District 02,* EPA Publication Date: March 5, 2003, p. 1.

605 Grand River Dam Authority, "Grand Lake O' the Cherokees," http://www.grda.com/water/ grand.htm (accessed July 30, 2006).

606 Jonna Polk, "Tar Creek is nation's largest Superfund site."

607 "Tar Creek (Ottawa County) Oklahoma, EPA ID# OKD980629844," EPA Region 6, *Congressional District 02,* EPA Publication Date: March 5, 2003, p. 1.

608 Ibid.

609 Julie DelCour, "The town that Jack built."

610 Nieberding, History of Ottawa County, p. 73.

611 Julie DelCour, "The town that Jack built."

612 David Averill, "Dodging the bullet: Abandoned mines ongoing hazard," Tulsa World, March 14, 2004, G-1.

613 Mike Jones, "That sinking feeling," Tulsa World, November 30, 2003, G-1.

614 Jonna Polk, "Tar Creek is nation's largest Superfund site."

615 Ed Keheley and Mary Ann Pritchard, Report to Governor Keating's Tar Creek *Superfund Task Force by the Subsidence Committee,* July 21, 2000, pp. 13–14.

616 Richard E. Meyer, "The Tar Creek Time Bomb," Los Angeles Times, http://www.science.uwaterloo.ca/earth/waton/s902.html (accessed April 16, 2003).

617 Nieberding, History of Ottawa County, p. 86.

618 Omer Gillham, "Collapsing into the past," Tulsa World, February 19, 2006, A-21.

619 David Averill, "Hellholes," Tulsa World, June 22, 2003, G-5.

620 Jim Myers, "Inhofe: U. S. 69 could cave in," Tulsa World, December 16, 2005, A-1.

621 Ed Keheley and Mary Ann Pritchard, Report to Governor Keating's Tar Creek *Superfund Task Force by the Subsidence Committee,* July 21, 2000, p. 14.

622 Kenneth V. Luza, "Stability problems associated with abandoned underground mines in the Picher Field, Northeastern Oklahoma," *Oklahoma Geological Survey Circular 88,* (Norman, Oklahoma, The University of Oklahoma, 1986), pp. 13–16.

623 Ed Keheley and Mary Ann Pritchard, Report to Governor Keating's Tar Creek *Superfund Task Force by the Subsidence Committee,* July 21, 2000, p. 22.

624 Brenda Luthy, "Kansas sinkhole continues to grow," Tulsa World, August 2, 2006, A-8.

625 Subsidence Evaluation Team, U.S. Army Corps of Engineers Tulsa District, Picher *Mining Field, Northeast Oklahoma Subsidence Evaluation Report,* January 2006, pp. 7.5–7.6.

626 Scott Christenson, "Contamination of Wells Completed in the Roubidoux Aquifer by Abandoned Zinc and Lead Mines, Ottawa County, Oklahoma," *U. S. Geological Survey Water-Resources Investigations Report 95–4150,* Abstract, Table 1, p. 6, http:// pubs.usgs.gov/wri/wri954150 (accessed January 16, 2006).

627 U. S. Geological Survey, "Project Title: Assessment of Ground-Water Flow and Recharge in the Boone Aquifer in Ottawa County, Oklahoma,," http://ok.water. usgs.gov/roj/boone.aquifer.html (accessed January 16, 2006).

628 EPA Superfund Record of Decision: Tar Creek (Ottawa County). EPA ID: OKD980629844. OU 01 Ottawa County, OK. June 6, 1984, #CSS Current Site Status.

629 Ibid.

630 Ibid., #AE Alternative Evaluations.

631 Ibid., #CSS Current Site Status.

632 EPA Region 6, Congressional District 02, "Tar Creek (Ottawa County) Oklahoma, EPA ID# OKD980629844," EPA Publication Date: January 10, 2006, p. 3.

633 Department of Environmental Quality, State of Oklahoma, "Tar Creek," Fact Sheets, http://www.deq.ok.us/1pdnew/FactSheet/Old Remediation Reports/ RemREp10–02%.pdf. (accessed January 26, 2006).

634 EPA Region 6, Congressional District 02, "Tar Creek (Ottawa County) Oklahoma, EPA ID# OKD980629844," EPA Publication Date: January 10, 2006, p. 1.

635 Agency for Toxic Substances and Disease Registry, Report to Congress—Tar Creek Superfund Site, *Ottawa County, Oklahoma,* http://.atsdr.cdc.gov/sites/ tarcreek/tarcreekreport-pl.html (accessed January 26, 2006).

636 Jonna Polk, "Tar Creek is nation's largest Superfund site."

637 EPA Region 6, Congressional District 02, "Tar Creek (Ottawa County) Oklahoma, EPA ID# OKD980629844," EPA Publication Date: January 10, 2006, pp 1–2.

638 Oklahoma Department of Environmental Quality, Oklahoma Plan for Tar Creek, May 2003, p. 31.

639 Ibid., p. 6.

640 Shaun Schafer, "Tar Creek buyout idea gets push," Tulsa World, May 3, 2003, A-1.

641 Jim Myers, "Tar Creek plan pushed," Tulsa World, May 5, 2003, A-1.

642 Archives, Tulsa World, http://www.tulsaworld.com/ArchiveSearch/Search/ SearchFormNew.asp (accessed February 23, 2006).

643 Janet Pearson, "Inhofe plan not the answer," Tulsa World, July 1, 2003, A-16.

644 "In Crisis," Tulsa World, February 21, 2005, A-19.

645 Jim Myers, "Inhofe to unveil Tar Creek Plan," Tulsa World, May 4, 2006, A-1.

646 Ken Neal, "Beyond dispute," Tulsa World, February 19, 2006, G-6.

647 Sara Plummer, "Dozens of Homes Lost," Tulsa World, May 13, 2008, 1.

The Quapaw—Into the Twenty-first Century

648 Matthew 24:13 (KJV)

649 Charles J. Kappler, comp. & ed., Indian Affairs: Laws & Treaties, Vol. 1 Laws, Compiled December 1, 1902, (Washington: Government Printing Office, 1904), pp. 566–567.

650 Baird, pp. 150–151.

651 Charles J. Kappler, comp. & ed., Indian Affairs: Laws & Treaties, Vol. 1. Laws, pp. 619–620.

652 Baird, The Quapaw Indian—A History of the Downstream People, pp. 150–151.

653 Ibid., pp. 152–154.

654 Ibid., pp. 155–156, 158–159.

655 Ibid., p. 160.

656 U.S. Supreme Court, U.S. v. Charles F. Noble, John M. Cooper, A. J. Thompson, A. S. Thompson, and V. E. Thompson , 237 U.S. 74 (1915), No. 127, http://westlaw. com/print/printstream.aspx?sv=Full&prft=HTMLE&mt=mt=Westlaw&f... (accessed December 11, 2006).

657 Baird, The Quapaw Indians—A History of the Downstream People, pp. 160–161.

658 Vern E. Thompson, " A History of the Quapaw," The Chronicles of Oklahoma, Vol. XXXIII, (Oklahoma City, Oklahoma: Oklahoma Historical Society, 1955), p. 360.

659 Baird, The Quapaw Indians—A History of the Downstream People, pp. 171, 203, 208.

660 Thompson, " A History of the Quapaw," p. 360

661 Baird, The Quapaw Indians—A History of the Downstream People, pp. 208–209.

662 Vern E. Thompson, Brief History of the Quapaw Tribe of Indians, (Pittsburg, Kansas: Mostly Books, 1994), Originally published in 1937.

663 Ibid., pp. 34–35.

664 Ibid., p. 34.

665 Baird, The Quapaw Indians—A History of the Downstream People, pp. 183–184, 186, 189.

666 Ibid., pp. 190–191.

667 Charles J. Kappler, comp. & ed., Indian Affairs: Laws & Treaties, Vol. III Laws, Compiled December 1, 1913, (Washington: Government Printing Office, 1913), p. 387.

668 Baird, The Quapaw Indians—A History of the Downstream People, pp. 156, 162–163.

669 U.S. Supreme Court, George W. Choate et.al., plffs. In err., v M. F. Trapp, Secretary of the State Board of Equalization, et.al., 224 U.S. 665 (1912), No. 809, http://caselaw.lp.findlaw.com/scripts/printer_friendly.pl?page=us/224/665.html (accessed August 18, 2006); Baird, *The Quapaw Indians—A History of the Downstream People*, p. 163.

670 Baird, The Quapaw Indian—A History of the Downstream People, pp. 172–173, 175.

671 Charles J. Kappler, comp. & ed., Indian Affairs: Laws & Treaties, Vol. IV. Laws, Compiled March 4, 1927, (Washington: Government Printing Office, 1904), p. 316.

672 Baird, The Quapaw Indians—A History of the Downstream People, p. 191.

673 Ibid., p. 205; Indian Affairs: Laws and Treaties, Vol. VI, Laws, Compiled from February 10, 1939 to January 13, 1971, (Washington: Government Printing Office, 1971), p. 1193.

674 Official Quapaw Website: "Government Organization Throughout Time," http:// geocities.com/Athens/Aegean/1388/gov.html?20072 (accessed February 2, 2007).

675 Charles J. Kappler, comp. & ed., Indian Affairs: Laws & Treaties, Vol. IV. Laws, p. 316.

676 Baird, The Quapaw Indians—A History of the Downstream People, pp. 212, 215; Indian Affairs: Laws *and Treaties*, Vol. VI, Laws, p. 1193.

677 S. E. Ruckman, "Quapaw tradition lives on, even in death," Tulsa World, May 11, 2005, A-9.

678 Baird, The Quapaw Indians—A History of the Downstream People, p. 202.

679 Department of Commerce, Bureau of the Census, Indian Population in the United States and Alaska *1910*, (Washington: Government Printing Office, 1915), pp. 20, 37.

680 Ruckman, "Quapaw tradition lives on, even in death," Tulsa World, A-9.

Epilogue—Journey's End

681 Proverbs 5:4–5 (KJV)

682 F. Jay Taylor, The United States and the Spanish Civil War, (New York: Bookman Associates, 1956), p.

19. Quote from Introduction by Claude G. Bowers.

683 Proverbs 23:23 (KJV)

684 Arnold, p. 143.

685 William J. Bennett, The Broken Hearth, (New York: Doubleday, 2001), pp. 174–175.

686 Ibid., p. 175.

687 Daniel Patrick Moynihan, http://www.infoplease.com/ce6/people/A0834298.html (accessed March 13, 2006).

688 "Moynihan on the family/Wilson on Marriage & Divorce," Interview of Daniel Patrick Moynihan by George Will on *This Week,* American Broadcasting Company, September 24, 2000, http:// archives.his.com/smartmarriages/2000-September/msg00047.html (accessed March 13, 2006).

689 "Gambling," Focus on Social Issues, http://www.family.org/cforum/fosi/gambling/gitus/ (accessed March 17, 2006).

690 National Indian Gaming Commission Tribal Gaming Revenues," National Indian Gaming Commission, http://www.nigc.gov/nigc/tribes/trigamrev2004to2003.jsp (accessed March 15, 2006).

691 Tony Batt, National Indian Gaming Report: Tribal gaming take: $19.4 billion, Los Vegas Review *Journal,* July 14, 2005, http://www.reviewjournal.com/lvrj_home/2005/Jul-14-Thu-2005/business/ 2501358.html (accessed March 15, 2006).

692 Problem Gambling.com, http://problemgamblinglcom/faq.html (accessed March 15, 2006).

693 Alexis de Tocqueville, Democracy in America, Revised Edition, Vol. II, Second Book, Chapter II, (London: The Colonial Press, 1900), p. 105.

694 Ibid., Volume II, Second Book, Chapter II, p. 104.

695 Ibid., Volume II, Second Book, Chapter IV, p. 109.

696 Ibid., Volume II, Second Book, Chapter V, p. 116.

697 Arnold, pp. 144–145.

698 Ibid., p. 63.

699 Tocqueville, Volume II, Second Book, Chapter VI, P. 119

700 Ibid., Volume II, Second Book, Chapter VI, p. 120.

701 Paul Starr, The Creation of the Media, (New York: Basic Books, 2004), p. 87.

702 Ibid., p. 86.

703 Ibid., pp. 396–397.

704 Ibid., pp. 398–399.

705 Ibid., p. 395.

706 Ibid., p. 9.

707 Alexis de Tocqueville, Democracy in America, The Henry Reeve Text as Revised by Francis Bowen, Volume I, Chapter XI, (New York, Alfred A Knoff, 1945), p. 188.

708 Ibid., Volume I, Chapter XI, p. 182.

709 Isaiah 60:22 (KJV)

Addendum—The North American Indian—Ancient Origins

710 Genesis 11:8 (KJV)

711 Earl H. Swanson, Warwick Bray, and Ian Farrington, The Ancient Americas, (New York: Peter Bedrick Books, 1989), p. 10.

712 Ibid., p. 1.

713 Bjorn Kurten, Before the Indians, (New York: Columbia University Press, 1988), p. 6.

714 Swanson, p. 2.

715 Ibid., p. 11.

716 Ibid., p. 2.

717 Kurten, p. 130.

718 Harold E. Driver, Indians of North America, Second Edition, Revised, (Chicago: The University of Chicago Press, 1969), p. 7.

719 Swanson, p. 2.

720 Charles C. Mann, 1491, (New York: Alfred A. Knopf, 2005) p. 15.

721 Ibid., p. 171.

722 Swanson, p. 11.

723 Ibid., pp. 72–73.

724 Shirley Gorenstein, ed., North America, (New York: St. Martin's Press, 1975), p. 75.

725 Ibid., p. 76.

726 George E. Hyde, Indians of the Woodlands—From Prehistoric Times to 1725, (Norman, Oklahoma: University of Oklahoma Press, 1962), pp. 3–6.

727 Swanson, pp. 72–73.

728 Hyde, p. 7.

729 Archaeologists have named the Adena after Adena estate in Chillicothe, Ohio, the location where the first big Adena mounds were excavated. (Hyde, pp.17–18.)

730 Hyde, pp. 17–18

731 This culture is named after Captain M. C. Hopewell. The first Mound Builder site was found on his farm on the Scioto River near Chillicothe, Ohio. (Hyde, p. 27.)

732 Hyde., p.33.

733 Ibid., p.37.

734 Ibid., pp. 8–10.

735 Ibid., p. 38–39.

736 Ibid., p. 60.

737 Gorenstein, p. 94.

738 Hyde, pp. 72–73.

739 Gorenstein, p. 101.

740 Edna Kenton, ed., The Indians of North America, Vol. 1, (New York: Harcourt, Brace & Company, 1927), pp. xiii-xv.

741 Hyde., pp.14–15.

742 Ibid., pp. 84–85.

743 Ibid., p. 107.

744 Francis Parkman, *The Jesuits in North America in the Seventeenth Century*, (Boston: Little, Brown, and Company, 1905), pp. 4–5

745 Ibid.

746 Hyde, pp. 85, 108.

747 Ibid., pp. 107–108.

748 Ibid., pp.86–87.

749 Ibid., pp. 91–92.

750 Richard B. Morris, ed., "Founding of the English Colonies, 1578–1732," Encyclopedia of American *History*, (New York: Harper & Brothers, 1953), p. 50.

751 Parkman, p. 262; Hyde., pp. 114–115.

752 Hyde., p. 101.

753 Kenton, The Indians of North America, Vol. 1, pp. 11–12.

754 Ibid., pp. 108, 110.

755 Ibid., p. 108.

756 Ibid., pp. 122–123.

757 Ibid., p. 163.

758 Parkman, pp. 268–271.

759 Hyde, p. 240.

760 C. Matthew Snipp, American Indians: The First of This Land, (New York: Russell Sage Foundation, 1989), p. 21.

761 Ibid., p. 15.

762 Ibid., p. 7.

763 Ibid., pp. 9–11.

764 Ibid., p. 14.

765 Henry R. Schoolcraft, History, Condition and Prospects of the Indian Tribes of the United States, (Philadelphia: Lippincott, Grambo & Company, 1853), pp.523–524.

766 Snipp, p. 13.

767 Ibid., pp. 29–30.

768 Morris S. Arnold, The Rumble of a Distant Drum, (Fayetteville, Arkansas: The University of Arkansas Press, 2000), p. 149.

769 Alvin M. Josephy, Jr., The Indian Heritage of America, (New York: Alfred A. Knopf, 1990), p. 28.

770 Ibid.

771 Hyde, pp. 164–165.

772 Ibid., p. 167.

INDEX

Abenaki Indians, 285

Able, A. W., 94

Abrams, Abner W., 89-92, 94, 121-122, 131, 252-255, 260-261

Adena, 292-293, 346

alchemy, 147

Algonquin, 33-34, 291, 296-298

American Federation of Labor (AFL), 223-225

Amerind, 285, 287

Andaste Indians, 296, 300

Archaic Culture, 289, 291

Arkansas (state), 23, 25, 27, 32, 42, 45, 49-54, 56-58, 60-62, 64-67, 72, 79, 88, 97-99, 102-103, 117, 190, 212, 252, 267, 277, 294, 302-303

Arkansas Gazette, 57, 277

Arkansas River, 25, 27, 29, 30, 32-34, 37, 41, 44, 50, 52, 54-58, 60, 64, 80, 88, 107, 123, 266, 274

Armer, Tom, 219, 220

Armstrong, Frank, 89

Arnold, Morris S., 42, 274

artillery, 141-144

Baird, W. David, 30, 77, 258-259

Baring & Company, 47

Barker, Ma, 190, 193-194

Barnett, Barney, 203-204

Barrow, Clyde, 190-192

Bartles, Jake, 118

Bartlesville, Oklahoma, 118-122

Baxter Springs, Kansas, 86, 126, 130, 134-136, 186, 209, 219-220, 227-228

Beinville, Jean Baptiste le Moyne de, 37

Bendelari, A. E., 138-139

Bennett, William, 267-268

Bering Strait, 115, 285, 287

Big Knife, Joseph, 94

Big Raft, 64

Blosser, Howard, 136

Blosser, Richard "Dick", 134-136

Blunt, General James G., 75-76, 78, 91, 95

Blytheville, Arkansas, 50-51

Blytheville, Missouri, 123

Boomer, 100, 102-103, 106-107, 109

Boone aquifer, 237, 239, 243-244

Booth, Melissa, 193-194

Bossu, Jean-Bernard, 30

Boudinot, Elias, 100, 102

Boudinot, Elias C., 100, 102

Bowers, Claude G., 264-265

Bowles, Cherokee Chief, 58, 69

Boyd, Police Chief Percy, 191-192

Brady, Roy A., 215-218

British, 36-38, 43-44, 48, 143-144

Brooks, Jeheil, 65

Brown, Thomas, 216-217, 219-220

Browne, Colonel Charles, 228-229

Bureau of Indian Affairs (BIA), 17, 19, 70, 74, 85-86, 93-94, 105, 246, 254, 260-261, 301

Burke, James, 13-14, 282

Caddo Indians, 36, 58, 60-56, 67-68, 70, 99, 101, 104, 112, 267, 296

Caddo Prairie, 62, 64

Calhoun, John C., 53-54, 60

Campbell, Constable Cal, 191-192

Campbell, David, 124-125

Campbell, Don, 124, 164

Campti, Louisiana, 64

Canadian River, 54, 58, 69-71

Carette, Father, 37

Carson, Former Oklahoma Congressman Brad, 247-250

Caruthersville, Missouri, 50

Catholic, 67, 80, 93, 95

Cayuga Indians, 105, 296

Central Mill, 182, 184-185, 187-188, 202, 218, 221, 235, 240

Chakchiuma Indians, 37

Charles III, 45

Charles IV, 45, 47

Cherokee Indians, 51-54, 58, 68-70, 79, 86, 88, 98, 101-102, 104-105, 110, 112, 115-116, 119-121, 126, 267

Cherokee County, Kansas, 128, 209, 227-229, 231, 235-236, 242

Cherokee Outlet, 100, 104, 107, 110, 113

Cherokee Phoenix, 100

Cheyenne Indians, 82, 101, 104, 110

Chickasaw Indians, 25, 32, 36, 45, 64, 70-71, 101, 105, 260

Choctaw Indians, 58, 68-70, 78, 101, 105, 260, 267

Choctaw Academy, 79

Civil War, 23, 70, 74-76, 81, 86, 90, 95, 99, 102, 106, 112, 125, 211

Clabber, Quapaw Chief Peter, 254-155, 260-261

Clark, William, 51-56

Cleveland, Grover, 86, 103, 106-107

Coburn, Oklahoma Senator Tom, 249-150

Comanche Indians, 70, 82, 99, 101, 104, 112

Commerce Mining and Royalty Company, 133, 181, 184

Committee for Industrial Organization (CIO), 224-225, 230-231

Comprehensive Environmental Response, Compensation, and Liability Act of 1980 (CERCLA), 236

Confederacy, 70-71

Confederate, 70-71, 74, 102, 125

Connell, D. L., 139,

Conway, Henry W., 60-61

Cooley, Commissioner of Indian Affairs D. N., 71

Cordilleran ice, 284

Couch, William L., 102-103, 109

Couture, Jean, 36

Cox, John, 123-125

Crawford Seminary, 80

Crawford, Samuel, 91-93, 95, 121, 253-254

Crawford, Commissioner of Indian Affairs Thomas H., 80

Crawford, Secretary of War William H., 52-53

Creek Indians, 70-71, 101, 103, 105, 114, 124, 236

Crittenden, Robert, 57, 60-61

Curtis, Vice President Charles, 256

Dawes Allotment Act of 1887, 92, 112

Dawes Commission, 113

Dawes, Henry L., 93

De Soto, Hernando, 34

Dhegiha Sioux Indians, 25-26-30, 32, 53, 293-295, 303

Desert Culture, 289

Devine, George, 190, 192

Downey, Mary Elzina, 116

Downstream People, 23, 25, 27, 36, 38, 62, 259-260, 266, 303

Drake, Colonel Edwin Drake, 118

Dry, Sheriff Eli, 218, 221-222, 227-228

Du Poisson, Father Paul, 37

Eagle Picher companies, 135, 138-139, 184-186, 202, 204, 214, 218, 221, 224-226, 228-229, 235, 245

Earlie, Ottawa Chief John, 103, 106

Eaton, Secretary of War John, 65-66, 79

Environmental Protection Agency (EPA), 15, 186, 232-233, 236-237, 240, 243-247

Erie Indians, 296, 300, 303

Eskimo-Aleut, 285, 287-289

Evans, Mike, 194-195, 218, 220-223, 227, 229-230

Ewart, Paul, 254-255

Exendine, Jasper, 118

Fish, Paschal, 88

Floyd, Pretty Boy, 190, 193

Ford, Tennessee Ernie, 173

Fort Carlos, 45, 49

Foucault, Father Nicholas, 37-38

France, 32, 35, 37-38, 44-45, 47, 49, 56, 98, 123, 143-144, 147, 268

French, 27, 29, 31-39, 41-43, 45, 47, 49, 55-56, 58, 80, 88, 98, 122-123, 142-143, 267, 274, 291, 294, 297, 303

Galena, Kansas, 129, 209, 228, 231, 242

George, Mrs. Maude, 193

Gibson, Arrell, 136, 258

Grand Lake of the Cherokees, 239

Grand River, 239

Gray, George, 65

Great Britain, 38, 43-44

Griffin, Quapaw Chief Victor, 256-257, 261

Griffith, Jo Ann, 186

Hadley, Lewis, 83-85, 88, 95

Hannum, Richard, 66

Harbaugh, M. D., 216-217

Harrison, President Benjamin, 106-107

Haskell, Charles N., 114-115

Hattonville, Oklahoma, 132-136

Hayworth, J. M., 86

Head, Colonel Ewell, 222-223, 227

Heckaton, Quapaw Chief, 54, 56-57, 60-61, 65-67, 69, 79-80

Hile, Louis, 20

Hockerville, Oklahoma, 19, 227-228, 230

hoisterman, 157, 163, 166, 172, 175, 195, 197, 199, 211

Hoover, J. Edgar, 193

Hopewell period, 292, 294

Horn, O. J., 199

Hotel, John, 83

Howard, W. A., 84

Hulsman, Ruth, 209-210

Hunker, John, 75

Huron Indians, 295-297, 299-300

Hyde, George, 299

Illinois Indians, 25, 27, 41, 295

Indian Knoll Culture, 290-292

Indian Removal Act of 1830, 99

Indian Territory, 23, 33, 58, 67-68, 74-75, 77, 81, 91, 94, 97-99, 103, 109, 113-115, 117-119, 121-122, 129, 132, 134, 190, 194, 252-253, 257, 267

Inhofe, Oklahoma Senator James, 242, 246-250

International Union of Mine, Mill and Smelter Workers, 213, 215, 219, 223-224

Iroquois Indians, 25, 41, 266, 295-297, 299-300, 303

Iroquois Confederacy, 297-298

Izard, Arkansas Territorial Governor George, 62, 65

Jefferson, Thomas, 43, 45, 47, 49, 52, 99

Jesuit Relations, 295, 298

Johnson, President Andrew, 76

Johnstone, William, 118-121

Jones, Hiram, 81-84, 130

Joplin, Missouri, 15, 123-128, 132, 135, 138, 152, 164, 180-181, 186, 192, 204, 207-208, 213-214

Joplin, Harris, 123

Journeycake, Chief, 118

Kansa Indians, 104, 112, 266, 293-294, 303

Kansas (state), 15-16, 18, 25, 52, 70-71, 74-75, 77, 80, 85-86, 88-90, 95, 99, 102-103, 107, 110, 112, 117-118, 123, 126-131, 134-135, 186, 190, 199, 206, 208-209, 212, 228, 230, 236, 242, 270, 303

Kansas National Guard, 228-229

Kappa (Quapaw village), 27, 33-34, 36

Karpis, Alvin, 193

Ka-she-ka, 75

Keeler, George B., 118-119

Keeler, W. W. "Bill", 120

Keller, Constable Ray, 227, 230

Ki-he-cah-te-da, 71

Kiowa Indians, 82, 101, 104, 112

Ku Klux Klan, 212

Landon, Kansas Governor Alf, 228-229

La Salle, Rene`-Robert Cavelier, Sieur de, 34-36, 45, 124

Laurentian Culture, 290-291

Laurentide ice, 284

Lewis, John L.   , 224-225

Lincoln, Abraham, 74

Lincolnville, Indian Territory, 122, 131, 134

Lippman, Walter, 278-279

Lookout, Osage Chief Fred and Julia, 121

Louisiana Purchase, 13, 23, 48-49, 54, 56, 98-99, 119, 267

Louis XIV, 35, 45, 123

Maledon, George, 98

Mann, Charles C., 41, 287

Mantle, Merlyn Johnson, 204

Mantle, Mickey, 117, 133, 202-205, 210

Mantle, Mutt, 202, 204-205

Marquette, Jacques, 32-34, 36, 38

Massachusett Indians, 296

Massasoit, 40-41

McConnell, Dave, 221-222

McKenzie, A. G., 89-90

McNaughton, John, 129-130

Methvin, Henry, 191-192

Miami Indians, 75, 82, 89, 132

Miami, Oklahoma, 15-16, 126, 131-132, 134, 137, 140, 193, 219-222, 244, 247, 251, 257

Miami Royalty Company, 132-133

Miller, General James, 56-57

Miller, Lavoice, 231

Mississippi River, 23-25, 27, 32-38, 41, 43-45, 47, 49-52, 54-56, 58, 61, 66, 122, 211, 266, 274, 294-296, 303

Mississippian Culture, 289, 294, 296

Missouri (state), 15, 25, 49-52, 66, 70, 74, 77, 80, 85, 99, 123-129, 132, 134, 140, 186, 190, 192, 206, 208, 212, 222, 229, 270, 293, 303

Mitchell, George, 74

Mitchigamea Indians, 33

Modoc Indians, 82-83, 87, 101, 105

Mohawk Indians, 296-297, 299-300

Monroe, James, 47, 56

Montagnai Indians, 295

Mooney, James, 301

Mound Builder, 289, 292-293, 295

Moynihan, Former New York Senator Daniel Patrick, 268

Munford, Dr. Morrison, 103

Murray, Oklahoma Governor "Alfalfa Bill", 114-115

Muskhogean Indians, 294

Na-Dene, 285

Napoleon, 45, 47-48

Narragansett Indians, 40, 296

Natchez Indians, 37, 294

Nation du Feu Indians, 297

National Industrial Recovery Act of 1933, 223

National Labor Relations Act of 1935 (Wagner Act), 224-225, 232

National Labor Relations Board, 224-226

National Priorities List (NPL), 232, 236-237, 242-243, 246

Neosho Indian Agency, 74, 76, 79

Neosho River, 77, 236, 239, 245

Neutral Indians, 296-297, 300

New Madrid earthquake, 50-51

New Orleans, Louisiana, 37, 38, 45, 47, 295

Nickles, Former Oklahoma Senator Don, 247, 249

Nigh, Oklahoma Governor George, 236

No Man's Land, 104, 112, 142

Nolan, Joe, 222

Northwest Ordinance of 1787, 98-99

Nuttall, Thomas, 27, 50-51

Ohio River, 23, 38, 98, 266, 291, 293-294, 303

Ohio Valley, 24-25, 34, 37-38, 41, 58, 96, 252, 292, 294,-295, 299, 303

Oklahoma National Guard, 222, 227

Oklahoma Plan, 245-250

Omaha Indians, 25, 27, 53, 80, 266, 293-294, 303

Oneida Indians, 296

Onondaga Indians, 296

Opothleyahola, 70-71

Osage Indians, 25, 52, 53, 70-71, 81-86, 88-89, 98-99, 101, 104, 112, 114, 118-121, 131, 266-267, 275, 293, 303

Osage Manual Labor School, 76, 80-81

Osborne, Charles B. (Uncle Charley), 117, 196-197

Osotouy (Quapaw village), 27, 35, 58

Otoe-Missouri Indians, 104, 112

Ottawa Indians, 71-86-87, 101, 103, 105, 137, 295, 297

Ottawa County, Oklahoma, 137, 193, 196, 209-210, 212, 218, 234, 236, 240-241, 243, 257

Ozark, 148-149, 212-213, 270, 275

Paleo-Indians, 285-287

Parker, Bonnie, 190-191

Parker, Judge Isaac, 97-98

Patterson, Rev. Samuel, 80, 95

Pawnee Indians, 101, 104, 296

Payne, Captain David L., 102-103, 109

Penacook Indians, 296

Peoria Indians, 75, 82, 86-87, 94, 101, 105, 122, 129-130, 137

Peoria Mining Company, 130

Pequot Indians, 296

Phillips, Frank, 119-121, 201-202

Picher, Oklahoma, 15-17, 19-20, 117, 126, 135-141, 146, 149, 169, 180, 184, 186-187, 204, 209, 213, 215, 220-222, 229-231, 235, 240-242, 244, 246-247, 249-251, 264

Picher Lead and Smelting Company of Joplin, 135-136

Picher Mining Museum, 20

Picher, O. H., 135

Picher, O. S., 135

Pilgrims, 40, 299

Pine Bluff, Arkansas, 64, 67

Platter, Clara, 199

Ponca Indians, 82-85, 100, 104, 112, 266, 293-294, 302

Pope, John, 65-67

Porter, Chief Pleasant, 114

Pyramids of Giza, 188-189

Quachita River, 51-52, 60

Quaker, 81-83, 95, 126

Quapaw, Oklahoma, 19

Quapaw Agency, 81, 90, 253

Quapaw Industrial Board School, 81

Quapaw, Charley, 84, 89, 256-257

Quapaw, John, 261

Ramsey, Nadine, 190, 192

Ramsey, Pascal "Pat", 190, 192

Ray, Abraham and Dardenne, 88

Red River, 54, 58, 60, 62, 64-65, 67-68, 113, 267, 294

Reign of Terror, 125

Richardson, Bobby, 205

Ridge, John, 100, 102

Ridge, Major, 100, 102

Robinson, Reid, 229-230

Rodgers, W. H., 217

Rogers, W. C., 114

Roosevelt, President Franklin, 214, 223, 261

Roosevelt, President Theodore, 115

Ross, John, 70, 100, 102

Roubidoux aquifer, 237, 239, 243-244

Sabo, George, III, 30-31

Samoset, 40

Sarasin, 58, 64, 67-68

Schermerhorn, Reverend J. F., 66

Schoenmaker, Jesuit Father, 80-81, 95

SchoolCraft, Henry, 301

Scott, Major General Winfield, 102

Seminole Indians, 70-71, 85, 101, 103, 105

Semple, Emma, 199-200

Seneca Indians, 66,68, 70, 75-76, 81, 87, 90, 101, 105, 129-130, 296, 299-300

Shawnee Indians, 66, 68, 70-71, 75-76, 87-88, 101, 104-105, 110, 129

Sidwell, Aubert, 192

silicosis, 207-209

Sinatra, Frank, 146

Siouans, 32, 291-294

Sioux Indians, 84, 295

Smithson, James, 154-155

Snow, George C., 74-76, 78-79, 95

Spain, 38, 44-45, 47, 49, 56

Spanish, 42, 44-45, 47, 49, 55, 72, 124, 129-130, 274

Sparks, O. W., 128

Spring River, 80, 129, 134, 239

Student Efforts Against Lead (S.E.A.L.), 17

Summers, J. V., 90-91

Superfund, 13, 15, 18, 186, 232-233, 236, 239, 241-248, 250-251, 276

Sutter, Johann, 123

Tallchief (Louis Angel), 82, 255, 261

Thompson, V. E. "Vern", 57, 255, 257-258

Tingle, William, 123-125, 164

Tobacco Nation Indians, 296, 299-300

Toby (mine mule), 170-171

Tocqueville, Alexis de, 271-274, 277-280, 282-283

Tongigua (Quapaw village), 27, 36

Tonti, Henri de, 34-36

Tourima (Quapaw village), 29, 36

Tousey, Benjamin, 90

Trail of Tears, 102, 116

Treaty of 1867, 75, 77, 81, 91

Treece, Kansas, 186, 209, 228, 230-231

Tribal Efforts Against Lead (T.E.A.L.), 17

Tri-State Mine, Mill and Smelter Workers Union (Blue Card Union), 223, 225-227, 229-230, 232

Tri-State Mining District, 117, 123, 126, 128, 136, 141, 149, 158, 166, 186, 189, 212-213, 229, 236, 258, 271

Tri-State Zinc-Lead Ore Producers' Association, 208

tuberculosis, 208-210

Tulsa World, 247-250

Tunica Indians, 25, 34, 41, 294

Tuttle, Emaline, 81

Vallier, Alphonsus, 88-89, 92

Vallier, Amos, 94

Vallier, Samuel, 75

Wahkonda, 31-32

Wampanoag, 40, 296

War-te-she, 69, 80

Washington, George, 38

Western Federation of Miners, 213

Whitebird, Robert, 262

Winnebago Indians, 298

Woodland Culture, 289-290

World War I, 137, 141-143, 145, 154, 222, 278-279

World War II, 259

Young, Ottawa County Sheriff Walter, 230-231

Printed in the USA
CPSIA information can be obtained
at www.ICGtesting.com
LVHW042054261023
762248LV00003B/31